# Developing Numeracy

## MENTAL MATHS

ACTIVITIES FOR THE DAILY MATHS LESSON

Hilary Koll and Steve Mills

A & C BLACK

# Contents

Reprinted 2007
First published 2004 by A & C Black Publishers Limited
38 Soho Square, London W1D 3HB
www.acblack.com

ISBN 978-0-7136-6911-4


Editors: Lynne Williamson and Marie Lister

The authors and publishers would like to thank Jane McNeill and Catherine Yemm for their advice in producing this series of books.

A CIP catalogue record for this book is available from the British Library.



Printed and bound in Great Britain by Cromwell Press Ltd, Trowbridge.

A & C Black uses paper produced with elemental chlorine-free pulp, harvested from managed sustainable forests.

# Introduction

**Developing Numeracy: Mental Maths** is a series of seven photocopiable activity books designed to be used during the daily maths lesson. This book focuses on the skills and concepts for mental maths outlined in the National Numeracy Strategy *Framework for teaching mathematics* for Year 2. The activities are intended to be used in the time allocated to pupil activities; they aim to reinforce the knowledge and develop the facts, skills and understanding explored during the main part of the lesson. They provide practice and consolidation of the objectives contained in the framework document.

## Mental Maths Year 2

To calculate mentally with confidence, it is necessary to understand the three main aspects of numeracy shown in the diagram below. These underpin the teaching of specific mental calculation strategies.

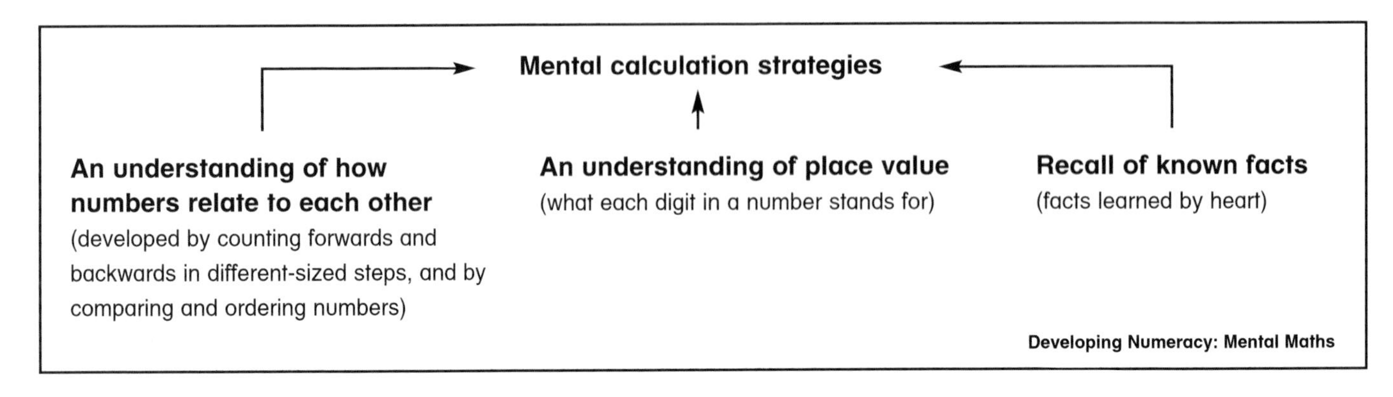


**Year 2** supports the teaching of mental maths by providing a series of activities which develop these essential skills. On the whole the activities are designed for children to work on independently, although this is not always possible and occasionally some children may need support.

**Year 2** develops concepts and skills for the different aspects of numeracy in the following ways:

### An understanding of how numbers relate to each other

- counting on and back from any number in ones, twos and tens;
- counting on and back in steps of three, four and five;
- recognising odd and even numbers;
- beginning to recognise two-digit multiples of 2;
- comparing and ordering numbers.

### An understanding of place value

- knowing what each digit in a two-digit number represents;
- partitioning two-digit numbers into tens and units;
- saying the number that is one or ten more/less than any number.

### Recall of known facts

Beginning to know by heart:

- all addition facts for each number to at least 10;
- all addition and subtraction facts for each number to at least 10, and the corresponding subtraction facts;
- all pairs of numbers with a total of 20;
- all pairs of multiples of 10 with a total of 100;
- multiplication facts for the 2, 5 and 10 times tables, and the corresponding division facts;
- doubles of all numbers to 15 and multiples of 5 to 50, and the corresponding halves.

### Mental calculation strategies

- putting the larger number first and counting on in tens and ones;
- beginning to add three single-digit numbers mentally (totals up to about 20);
- partitioning into '5 and a bit' or tens and units when adding, then recombining;
- finding a small difference by counting up;
- identifying near doubles, using doubles already known;
- adding or subtracting 9, 11, 19 or 21 by adding or subtracting 10 or 20, then adjusting by 1;
- using patterns of similar calculations;
- using known number facts and place value to add or subtract pairs of numbers mentally;
- multiplying a single-digit number by 1, 2, 3, 4, 5 and 10.

## Extension

Many of the activity sheets end with a challenge (**Now try this!**) which reinforces and extends the children's learning, and provides the teacher with an opportunity for assessment. On occasion it may be helpful to read the instructions with the children before they begin the activity. For some of the challenges the children will need to record their answers on a separate piece of paper.

## Organisation

Very little equipment is needed, but it will be useful to have the following resources available: coloured pencils, counters, cubes, dice, scissors, coins, squared paper, number lines and number tracks.

To help teachers select appropriate learning experiences for the children, the activities are grouped into sections within the book. However, the activities are not expected to be used in this order; the sheets are intended to support, rather than direct, the teacher's planning.

Some activities can be made easier or more challenging by masking or substituting numbers. You may wish to re-use pages by copying them onto card and laminating them.

## Teachers' notes

Brief notes are provided at the foot of each page giving ideas and suggestions for maximising the effectiveness of the activity sheets. These can be masked before copying.

## Whole-class warm-up activities

The following activities provide some practical ideas which can be used to introduce the main teaching part of the lesson.

### Think of a number

Play this game by asking questions such as:

*I'm thinking of a number. When I add 11 to it the answer is 19. What is my number?*

*I'm thinking of a number. When I take away 8 the answer is 12. What is my number?*

### What's the question?

Write a number on the board, such as 19, and explain that this is the answer to a question. Ask the children: *Can you give me a question with the answer 19?* (for example, 12 + 7); *Can anyone give me another question with the answer 19?; Can you think of a subtraction question with the answer 19?* Explore the range of calculations that give the answer 19 and discuss the patterns that emerge: for example, 18 + 1, 17 + 2, 16 + 3 and 20 – 1, 21 – 2, 22 – 3.

### Number cards

Give each child a set of 0 to 20 cards, which should be spread randomly face up on the table. Call out a number and ask: *Can you make the number that is 9 (or 11) more/less than my number?*

Call out a number (for example, 23) and ask: *Can you hold up two numbers that add to make my number?* Explore the different sums that are produced.

### Missing numbers

Write some number statements on the board, using a box in the place of one of the numbers in the question: for example, 8 + ☐ = 17, ☐ + 7 = 18. Invite the children to come and fill in the missing numbers.

### Fast facts

Call out addition questions such as *12 add 6 equals...?* and *9 plus 11 makes...?* Repeat with subtraction questions, using a range of appropriate vocabulary.

### Poster numbers

Write some numbers on the board or on a poster. Call out questions (for example, *9 plus 6*; *25 minus 8*) and ask the children to find the answers on the poster.

| | | | |
|---|---|---|---|
| 20 | | 17 | |
| | 15 | | 6 |
| 2 | | 12 | |
| | 8 | | 35 |
| 32 | | 40 | |

Then choose two numbers from the poster and ask the children to find the difference between them.

Say a number and ask the children to use two numbers from the poster in an addition or a subtraction to make that number: for example, if you say the number 9, the children would use the subtraction 17 – 8.

# Campfire counting

**These five Beaver Scouts count around in** ones **up to 100. Tom starts with zero.**

**1.** Write all the numbers Ravi will say.

3, ______________________________

______________________________

**2.** Write all the numbers Jack will say.

1, ______________________________

______________________________

**3.** Write all the numbers Tom will say.

0, ______________________________

______________________________

**Now Tom says 100. They** count back **to zero.**

- **Write all the numbers Ravi will say.**

97, ______________________________

______________________________

**Teachers' note** Encourage the children to count around the circle in the picture, pointing to each child in turn. Ask them what they notice about the number patterns. The instructions could be changed for more confident children: for example, they can be asked to count around in twos.

# Two-whit two-whoo!

- Cut out the cards.
- How many eyes? Count in twos.

  Write the number on the back.

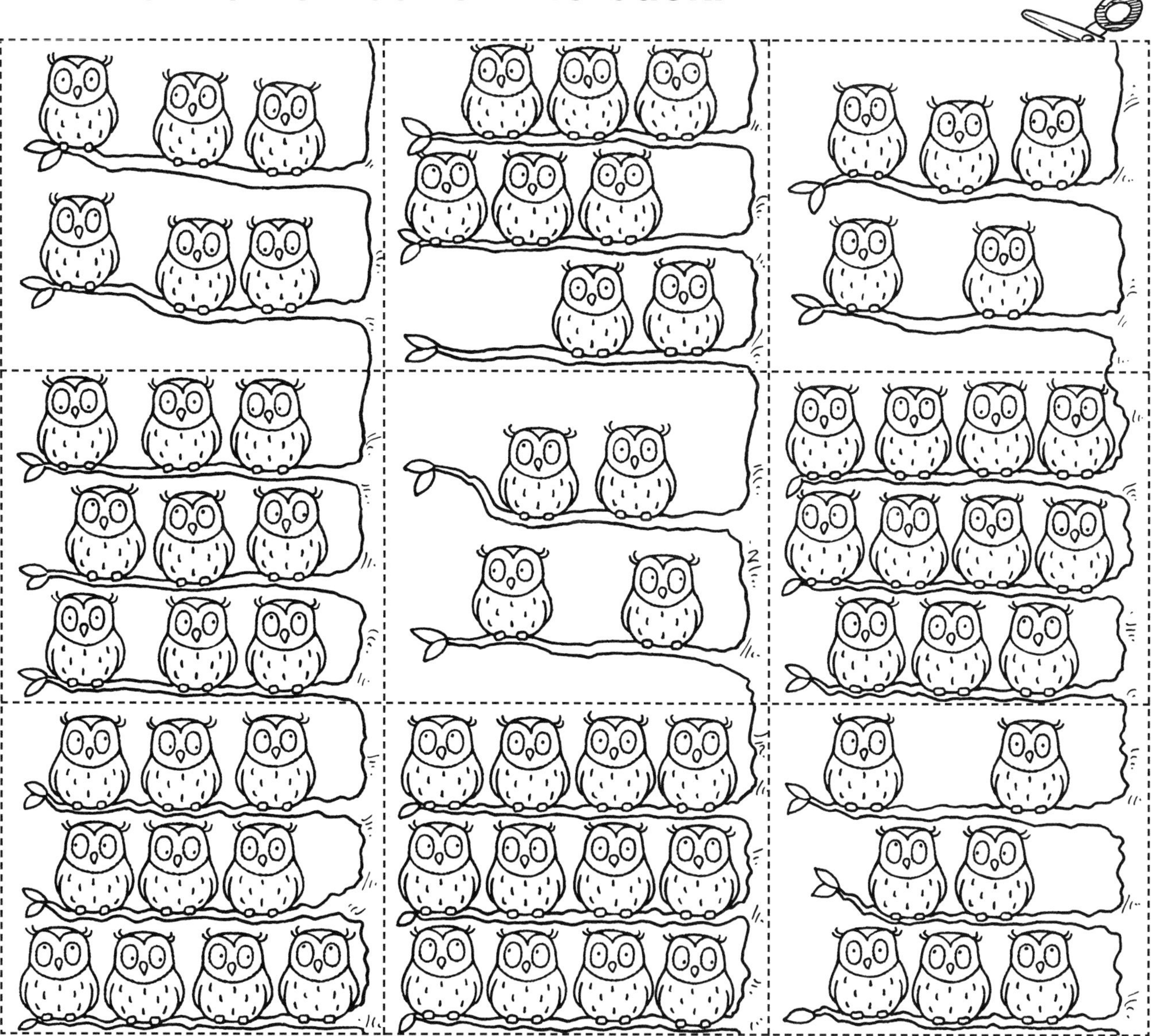

- The numbers are all even. Put them in order.

- Ring the even numbers.

17 16 18 21 20 14 13 19 12 22 15

37 26 28 31 30 24 23 49 32 42 25

**Teachers' note** Demonstrate jumping in twos along a number line from and back to zero. Encourage the children to realise that when counting in twos from zero the numbers are even and are multiples of 2. At the end of the lesson, draw attention to the units digits of even numbers and show that counting in twos starting on an odd number produces a sequence of odd numbers.

# Counting fun

**These ten Brownies count around in** ones **up to 100. Alex starts with zero.**

1. Write all the numbers Ruby will say.
   2, ______

2. Write all the numbers Leela will say.
   6, ______

3. Write all the numbers Lucy will say.
   8, ______

- **Count back in tens from these numbers.**

94, 84, ______

97, ______

99, ______

**Teachers' note** Discuss with the children that every tenth number will be counting on in tens. This could be shown on a number line/track or a 100-square. Invite the children to explain the patterns they notice when counting on in tens. During the plenary, ask all the children to read the solutions aloud, to reinforce counting on and back in tens.

# Sequence puzzle

- **Find the circled number in the puzzle. Follow the instructions.**

Follow the direction of the arrow.

| | | |
|---|---|---|
| ~~(1) Count on in 2s~~ | (2) Count back in 2s | (3) Count on in 3s |
| (4) Count on in 1s | (5) Count on in 10s | (6) Count on in 5s |
| (7) Count on in 3s | (8) Count back in 1s | (9) Count on in 10s |
| (10) Count on in 1s | (11) Count on in 4s | (12) Count on in 2s |
| (13) Count on in 10s | (14) Count on in 5s | |

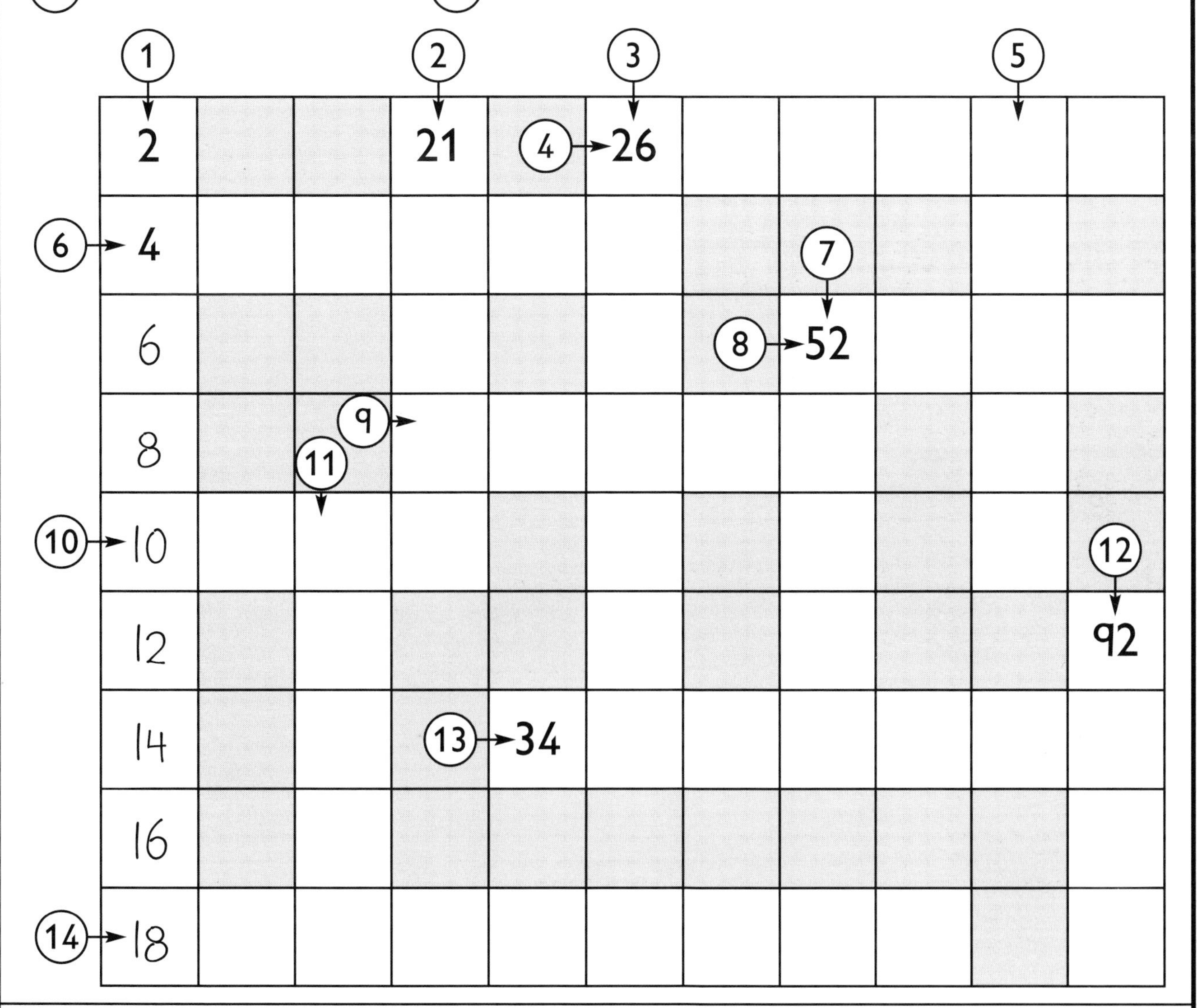

- **Make up another puzzle like this. Use squared paper.**

**Teachers' note** Before beginning, practise counting on and back in twos, threes, fours, and so on. Encourage the children to cross off each instruction as they go. Point out that some instructions involve counting on and some involve counting back. The children may find number lines or 100-squares helpful. They will need squared paper for the extension activity.

# Number split game

## • Play this game with a partner.

**You need:**
- cubes in two colours
- a set of 0–100 number cards, with the numbers 0–9 and the multiples of 10 removed

☆ Shuffle the cards and put them in a pile, face down.

☆ Take turns to pick a card.
Split the number into two parts, like this: 27 = 20 + 7

☆ If you can find both parts on the grid, cover each part with a cube in your colour.

☆ The winner is the player with the most cubes on the grid when all the numbers are covered.

| | | | | | |
|---|---|---|---|---|---|
| 20 | 90 | 60 | 30 | 10 | 90 |
| 40 | 70 | 80 | 50 | 30 | 70 |
| 10 | 80 | 40 | 20 | 60 | 50 |
| 1 | 9 | 2 | 8 | 5 | 6 |
| 2 | 7 | 4 | 7 | 3 | 1 |
| 8 | 9 | 5 | 3 | 6 | 4 |

**Teachers' note** Each pair needs one copy of this sheet. As an extension, write on the board three multiples of 10 (for example, 20, 60, 40) and three single digits (for example, 5, 9, 1), and ask the children to combine the tens and units to make as many two-digit numbers as they can.

# Fried eggs

- **Tick the larger number in each pair.**
- **Write a number that lies between them.**

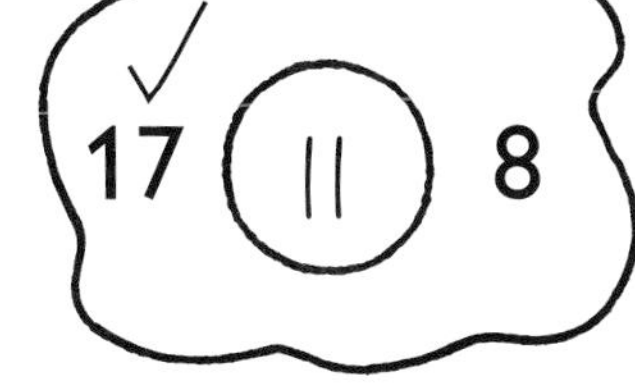

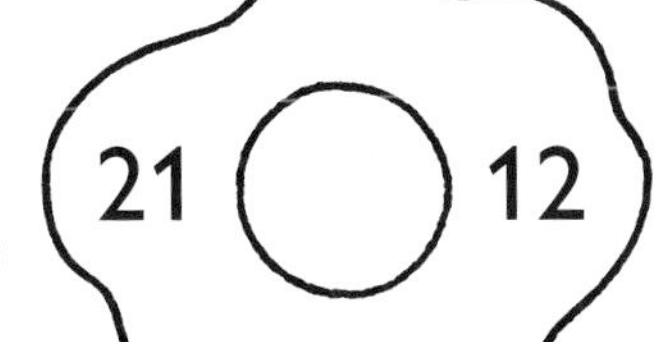

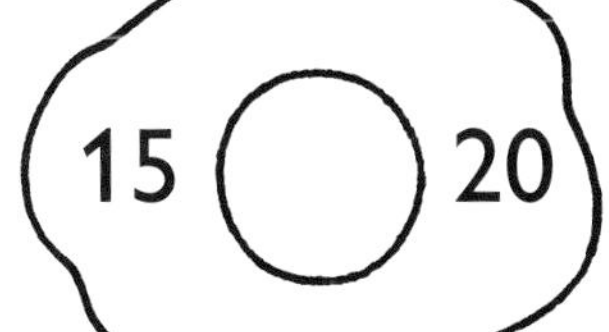

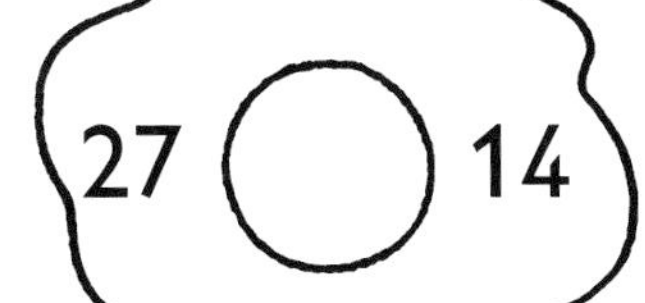

| | | | |
|---|---|---|---|
| 17 ✓ ( 11 ) 8 | 21 ( ) 12 | 19 ( ) 27 | 15 ( ) 20 |
| 27 ( ) 14 | 32 ( ) 29 | 16 ( ) 51 | 66 ( ) 69 |
| 34 ( ) 40 | 28 ( ) 82 | 33 ( ) 25 | 50 ( ) 47 |
| 28 ( ) 36 | 54 ( ) 45 | 38 ( ) 45 | 93 ( ) 99 |
| 72 ( ) 84 | 67 ( ) 56 | 87 ( ) 92 | 68 ( ) 71 |

- **Write these numbers in order.**

**Start with the smallest.**

30, 49, 96, 41, 77, 64, 93, 69

______________________________

**Teachers' note** Some children may need a number line/track or 100-square. Invite the children to say their answers aloud: for example, 'Twenty-one is larger than twelve, and eighteen lies between them.' Discuss the range of possible answers for different questions and ask the children to say which pairs of numbers in the questions have the fewest whole numbers lying between them.

# Number chains

- **Write what is added or subtracted each time.**

- **Complete this number chain.**

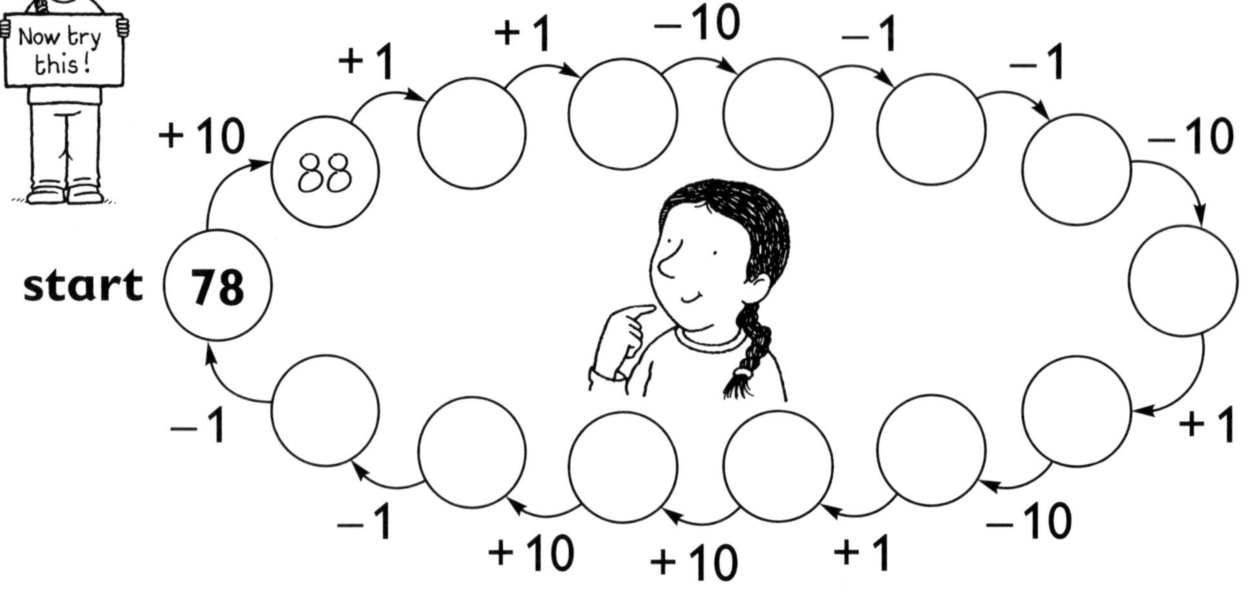


**Teachers' note** During the first part of the lesson, revise adding and subtracting 10 to/from any two-digit number. The children could be asked to write numbers on the board to form a similar chain as you give instructions such as 'add one', 'plus ten' 'subtract one', 'take ten'. During the plenary, encourage the children to read out the numbers and the additions/subtractions in the chains.

# Fun at the fair

• **Fill in the missing numbers.**

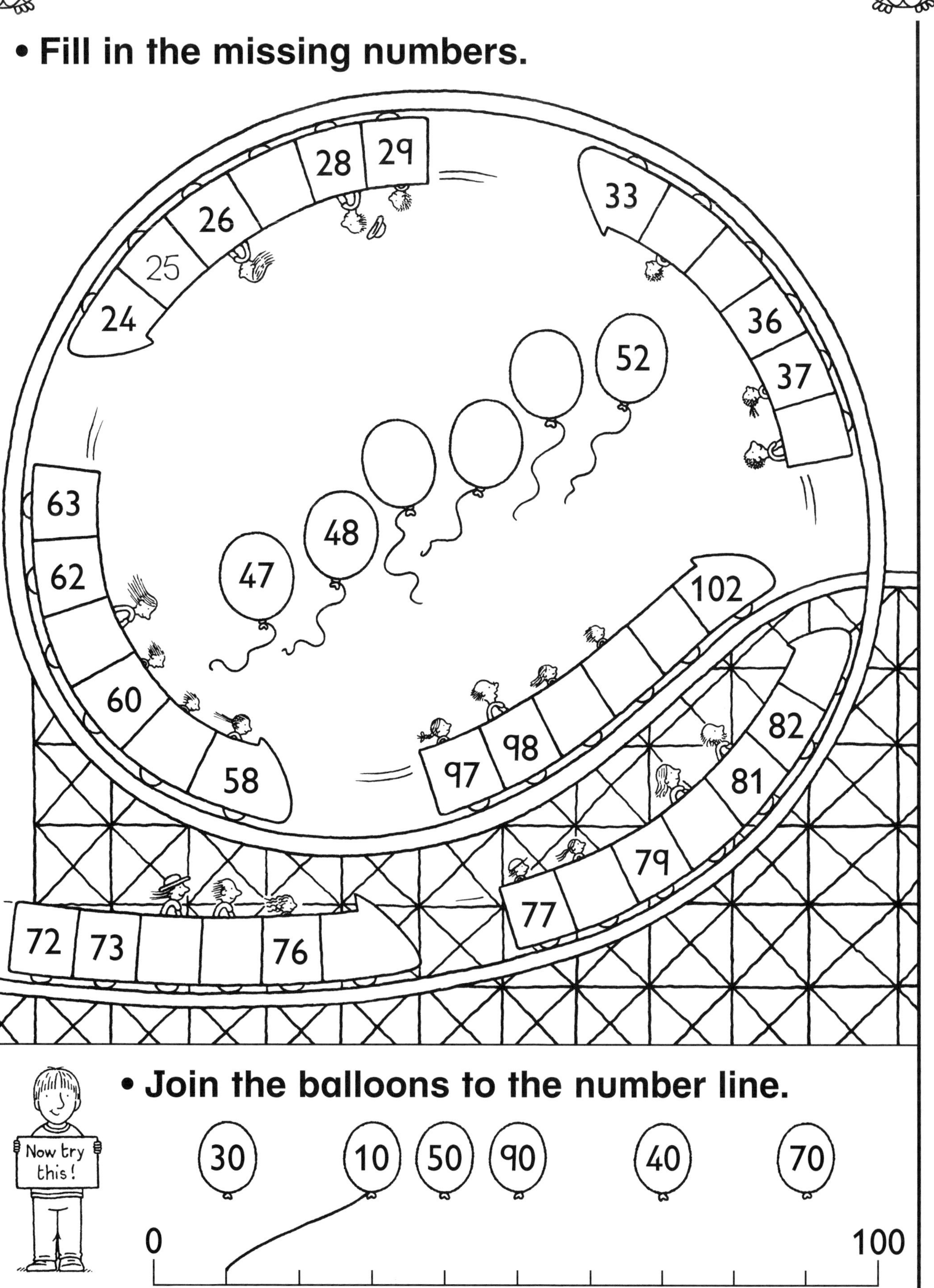


• **Join the balloons to the number line.**

30 10 50 90 40 70

0 100

Now try this!

**Teachers' note** In order to calculate mentally using strategies such as counting on or back, bridging through 10 or a multiple of 10 or 100, and so on, it is vital that the children have an understanding of where numbers are in relation to each other. Watch out for errors such as saying that 70 lies between 79 and 81, or that the number that follows 100 is 200.

# Switch-it witch

- A witch has switched ten pairs of numbers in the grid.
- Ring each pair of switched numbers.

| 1 | 2 | 3 | 4 | 5 | 6 | 7 | 8 | 9 | 10 |
|---|---|---|---|---|---|---|---|---|---|
| 11 | (13 | 12) | 14 | 15 | 16 | 17 | 18 | 20 | 19 |
| 21 | 22 | 23 | 24 | 35 | 26 | 27 | 28 | 29 | 30 |
| 31 | 32 | 33 | 34 | 25 | 36 | 37 | 38 | 49 | 40 |
| 41 | 42 | 52 | 44 | 45 | 46 | 47 | 48 | 39 | 50 |
| 51 | 43 | 53 | 54 | 56 | 55 | 57 | 58 | 59 | 60 |
| 61 | 62 | 63 | 64 | 65 | 66 | 67 | 68 | 69 | 80 |
| 71 | 81 | 73 | 74 | 75 | 76 | 77 | 78 | 79 | 70 |
| 72 | 82 | 83 | 84 | 86 | 85 | 87 | 99 | 89 | 90 |
| 91 | 92 | 93 | 94 | 95 | 96 | 97 | 98 | 88 | 100 |

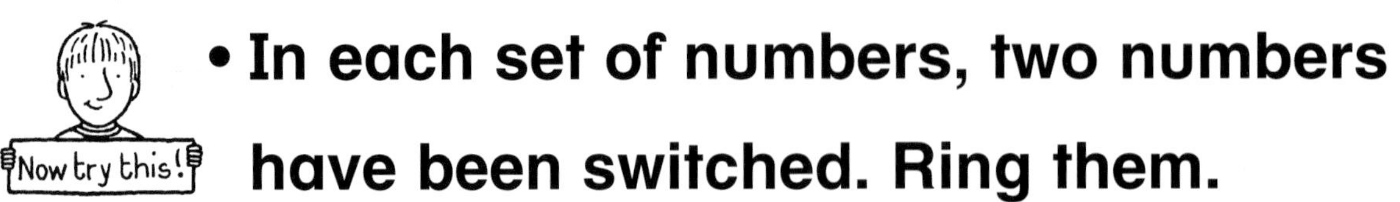

- In each set of numbers, two numbers have been switched. Ring them.

13 (25) 22 (18) 29

37 44 45 54 53

27 72 35 28 86 95

52 58 88 96 95 99

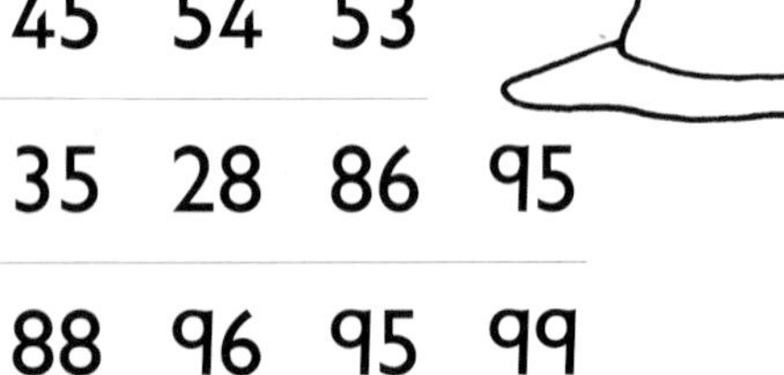

18 17 19 20 24

43 52 85 76 95

66 77 69 75 76 68

34 81 87 86 82 88

**Teachers' note** If you have a magnetic 100-square, or a 100-square made of individual cards, you could do a similar activity as a mental/oral starter. Switch pairs of numbers and invite the children to identify them. When you return the numbers to the correct places, ask the children to read the numbers aloud (this will help them to check that the order is now correct).

# Be a winner!

- **Write any numbers from 1 to 10 on the balls.**
- **Follow these instructions for each game.**

☆ Your teacher will write two numbers on the board. Find their total. If you have a ball with that number, cross it off.

☆ The winner is the first to cross off all six balls.

| | | | | | | |
|---|---|---|---|---|---|---|
| Game 1 | ◯ | ◯ | ◯ | ◯ | ◯ | ◯ |
| Game 2 | ◯ | ◯ | ◯ | ◯ | ◯ | ◯ |
| Game 3 | ◯ | ◯ | ◯ | ◯ | ◯ | ◯ |
| Game 4 | ◯ | ◯ | ◯ | ◯ | ◯ | ◯ |
| Game 5 | ◯ | ◯ | ◯ | ◯ | ◯ | ◯ |
| Game 6 | ◯ | ◯ | ◯ | ◯ | ◯ | ◯ |
| Game 7 | ◯ | ◯ | ◯ | ◯ | ◯ | ◯ |
| Game 8 | ◯ | ◯ | ◯ | ◯ | ◯ | ◯ |

**Teachers' note** This is a whole-class activity. Once the children have written numbers on all their balls, begin Game 1 by writing pairs of numbers on the board with a total of 10 or less (include zero in some pairs). Reinforce addition vocabulary by asking, 'What is … plus …?'; 'What is the sum/total?' The number range can be changed to make the additions harder, or subtraction can be used.

# Chocolate facts

**These bars of chocolate are broken in two.**

- **Each bar has 10 chunks. Write four facts.**

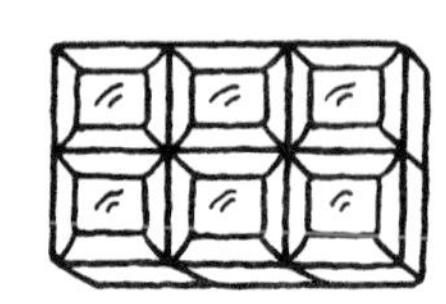
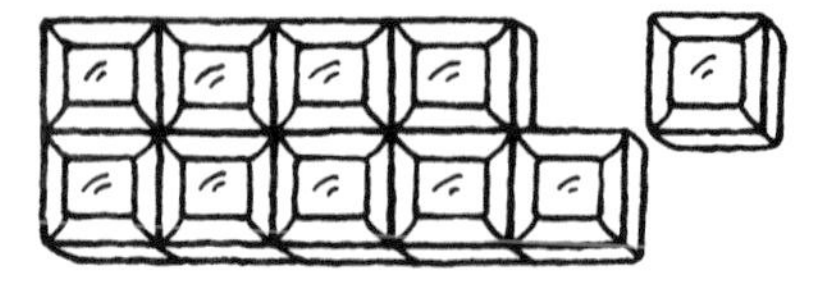

4 + 6 = 10
6 + 4 = 10
10 − 4 = 6
10 − 6 = 4

- **Each bar has 9 chunks. Write four facts.**

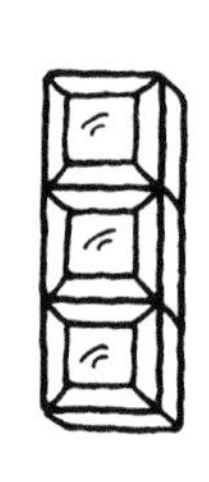
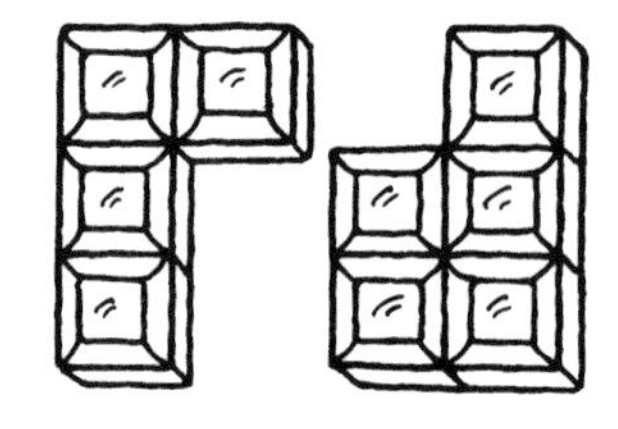
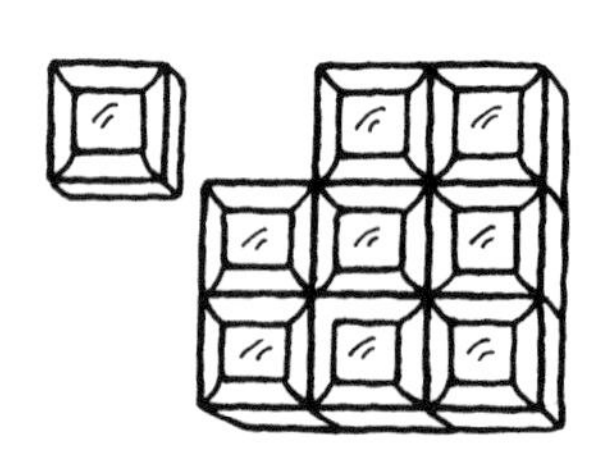

6 + 3 = 9

- **Complete these as quickly as you can.**

| | | |
|---|---|---|
| 3 + ☐ = 10 | ☐ + 6 = 9 | 10 − ☐ = 4 |
| 8 + ☐ = 9 | ☐ − 2 = 8 | 9 − ☐ = 4 |
| 1 + ☐ = 10 | ☐ + 6 = 10 | 9 − ☐ = 1 |
| 2 + ☐ = 9 | ☐ − 1 = 9 | 10 − ☐ = 7 |

**Teachers' note** For the extension activity, you could ask the children to cover or fold the sheet to hide the facts above. As a further extension, ask the children to draw different ways of splitting chocolate bars with six, seven or eight chunks, and to write the related number facts. They should be encouraged to learn the facts by heart.

# Total teaser

**Each child thinks of two numbers which total 20.**

- **Write the missing numbers.**
- **Then check your answers. Add the numbers together.**

- **Complete these as quickly as you can.**

20 – 17 = ☐ 20 – 6 = ☐ 20 – 12 = ☐

20 – ☐ = 5 20 – ☐ = 1 20 – ☐ = 13

**Teachers' note** Encourage the children to use their knowledge of number pairs with a total of 10 (for example, 4 + 6 = 10 so 14 + 6 = 20 and 4 + 16 = 20). Remind them that the order in which numbers are added is not important. For the extension activity, you could ask the children to cover or fold the sheet to hide the facts above.

# Cats and dogs

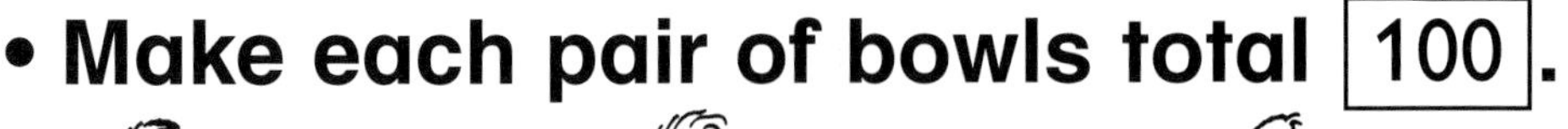

- **Make each pair of bowls total 100.**

- **Join pairs of dogs which total 100.**

- **Write which of these numbers total 100.**

Example:
85 + 15 = 100

| 25 | 5 | 65 | 35 | 75 | 55 | ~~15~~ | ~~85~~ | 45 | 95 |
|---|---|---|---|---|---|---|---|---|---|

**Teachers' note** Encourage the children to make the link between related facts: for example, 2 + 8 = 10 so 20 + 80 = 100. Make sure that they have an opportunity to say these number pairs aloud, using a range of vocabulary: for example, 'The total/sum of 20 and 80 is 100'; '20 plus 80 equals 100'; 'If I add 20 to 80 I get 100'; '20 and 80 more is 100'.

# Cross-number puzzles

- **Solve these puzzles. When you add, count on from the larger number.**
- **Then make up the missing clues.**

| | | | | |
|---|---|---|---|---|
| 1 2 6 | 6 | | 2 | |
| | | 3 | | |
| | 4 | | | 5 |
| 6 | | | 7 | |
| | | 8 | | |

**Across**
1 8 + 18
2 7 + 59
3 3 + 46
4 4 + 9
6 ______
7 ______
8 8 + 75

**Down**
1 8 + 20
2 ______
3 ______
4 9 + 3
5 6 + 89
6 6 + 68
7 2 + 61

**Across**
1 3 + 39
2 9 + 49
3 7 + 24
4 6 + 58
6 ______
7 ______
8 7 + 17

**Down**
1 7 + 40
2 ______
3 ______
4 6 + 63
5 8 + 67
6 9 + 74
7 6 + 48

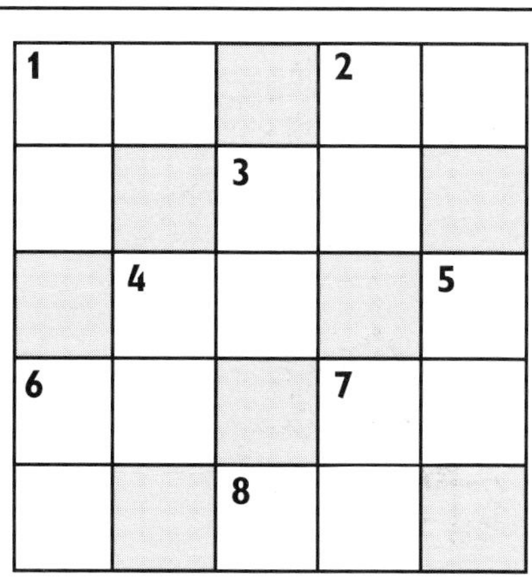

| | | | | |
|---|---|---|---|---|
| 1 | | | 2 | |
| | | 3 | | |
| | 4 | | | 5 |
| 6 | | | 7 | |
| | | 8 | | |

**Across**
1 7 + 88
2 7 + 57
3 4 + 59
4 7 + 51
6 ______
7 ______
8 5 + 68

**Down**
1 3 + 89
2 ______
3 ______
4 8 + 46
5 7 + 84
6 8 + 64
7 4 + 79

- **Copy the grid above onto squared paper.**
- **Make up your own cross-number puzzle.**

**Teachers' note** Check that the children know how the clues for a cross-number puzzle work. Remind them that the order of addition does not matter (for example, 7 + 88 is the same as 88 + 7), and that it is easier to count on from the larger number. The children may find number lines or 100-squares helpful. They will need squared paper for the extension activity.

# Wagon wheels game

## • Play this game with a partner.

☆ Place a counter on 'start' on each wheel.

☆ Take turns to roll the dice. Move each counter that number of places.

☆ Write the three numbers you land on and find the total. **Example:** 5 + 4 + 5 = 14

☆ The player with the higher total wins.

**You need:**
- a dice
- six small counters
- a copy of this sheet each

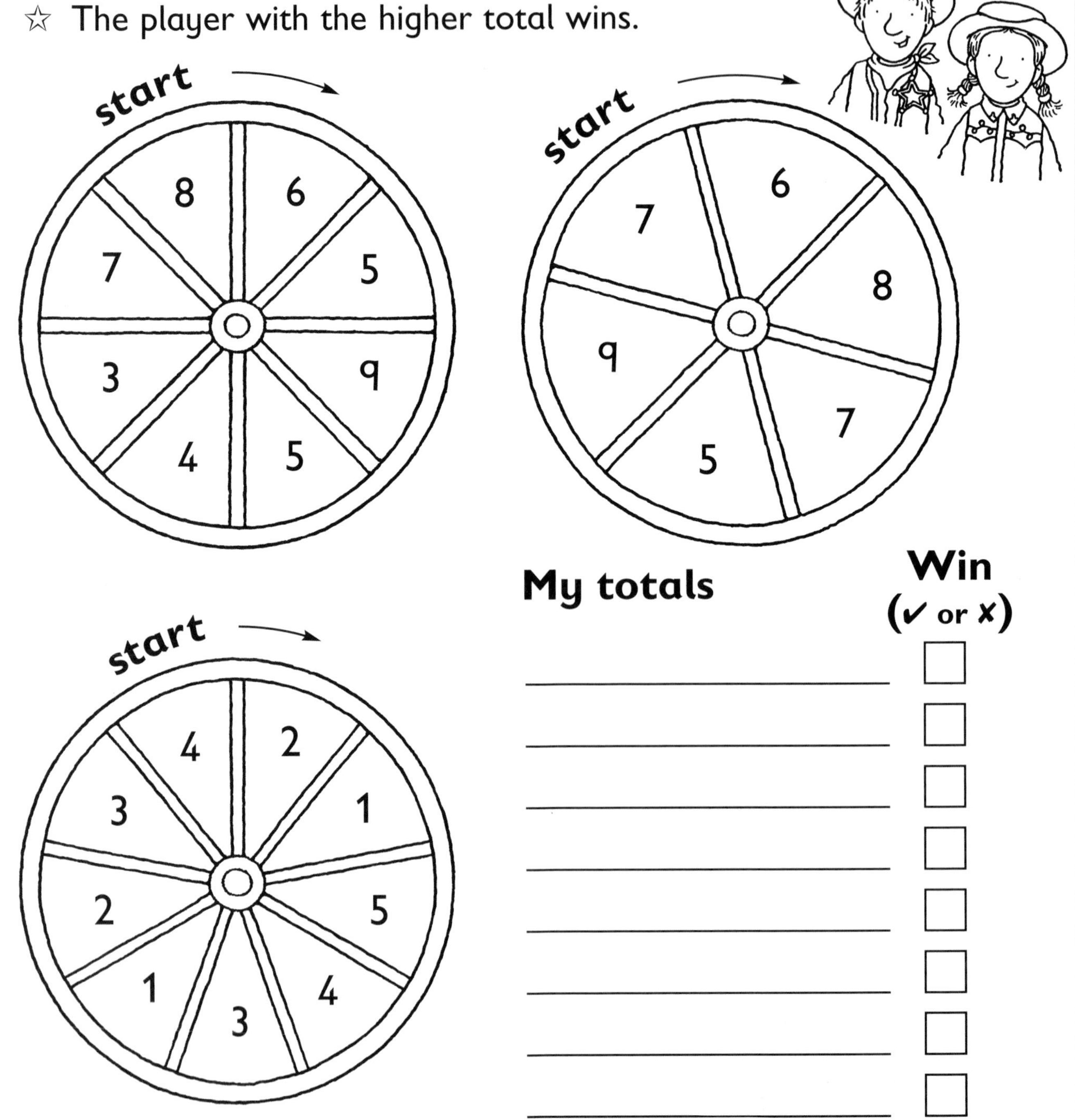




**Teachers' note** Each child needs a copy of this sheet. First discuss strategies for adding several numbers, such as counting on from the largest number or using doubles. Encourage the children to discuss their strategies with a partner and to read the questions and answers aloud, using a range of vocabulary. They can check their answers by adding the numbers in a different order.

# Whose balloons?

- **Join each child with the correct balloon.**

18 29 27

| 5 + 3 | 5 + 2 | 15 + 3 | 15 + 1 | 15 + 4 | 25 + 2 | 25 + 4 |
|---|---|---|---|---|---|---|

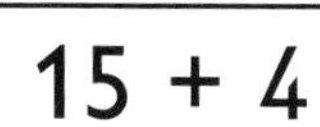

36

| 5 + 1 | 15 + 2 | 5 + 4 | 25 + 1 | 25 + 3 | 35 + 2 | 35 + 1 |
|---|---|---|---|---|---|---|

- **Use your answers above to complete these.**

8 + 6 = [5 + 3] + [5 + 1] = 10 + 4 = 14

7 + 16 = [ ] + [ ] = ____

18 + 17 = [ ] + [ ] = ____

27 + 16 = [ ] + [ ] = ____

- **Choose six other pairs of balloons.**
  **Find the totals.**

**Teachers' note** This activity introduces partitioning numbers into '5 and a bit', '15 and a bit', '25 and a bit', and so on, as a means of adding numbers with the unit digits 6, 7, 8 and 9. Revise additions such as 15 + 5, 15 + 15, 25 + 15, 25 + 25 before beginning the activity.

# T-shirt totals

- **Split the numbers into** multiples of 5 **and a bit.**
- **Fill in the missing numbers to find the total.**

16 + 7          18 + 6

17 + 5          17 + 7

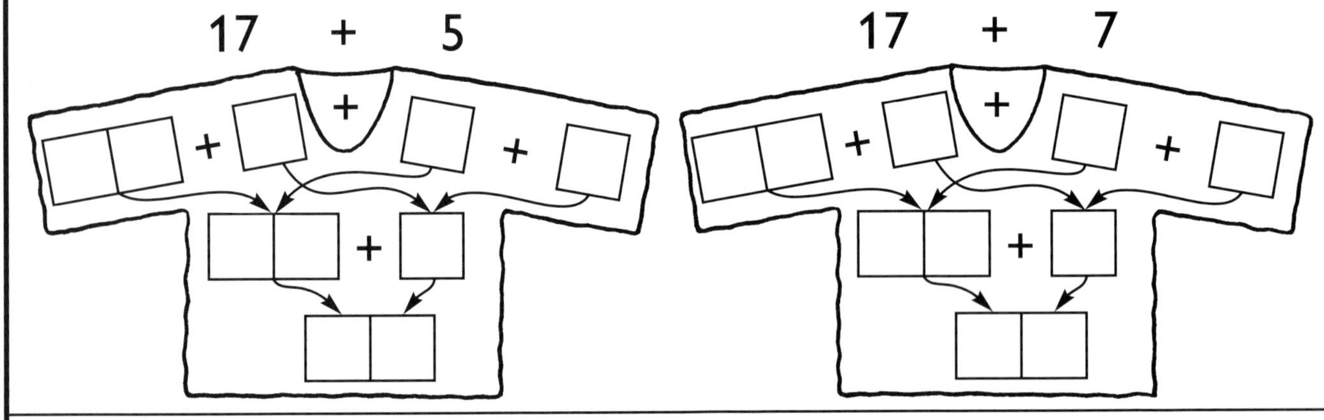


38 + 6          28 + 9

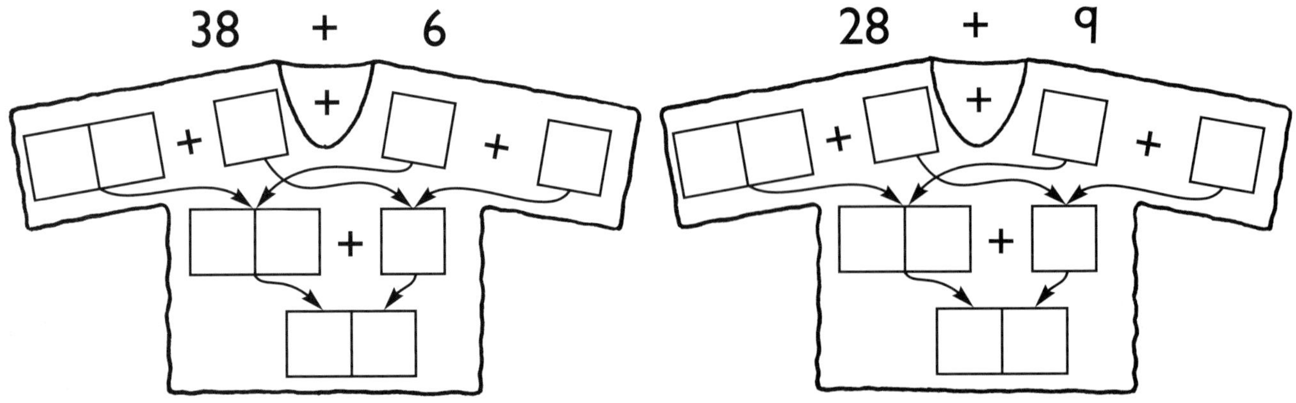

- **Add these in your head in the same way.**

18 + 25 = ☐     17 + 22 = ☐     28 + 26 = ☐

28 + 29 = ☐     35 + 17 = ☐     18 + 37 = ☐

36 + 27 = ☐     28 + 38 = ☐     37 + 36 = ☐

**Teachers' note** The children may benefit from completing the activity on page 21 before tackling this one. Revise additions such as 15 + 5, 15 + 15, 25 + 15, 25 + 25 before beginning the activity. When partitioning, encourage the children to find the largest multiple of 5 that can be taken from the number, so that the 'bit' is less than 5 (for example, 27 is 25 + 2, rather than 20 + 7).

# Partition addition

- **Use these jewels to help you add pairs of two-digit numbers.**

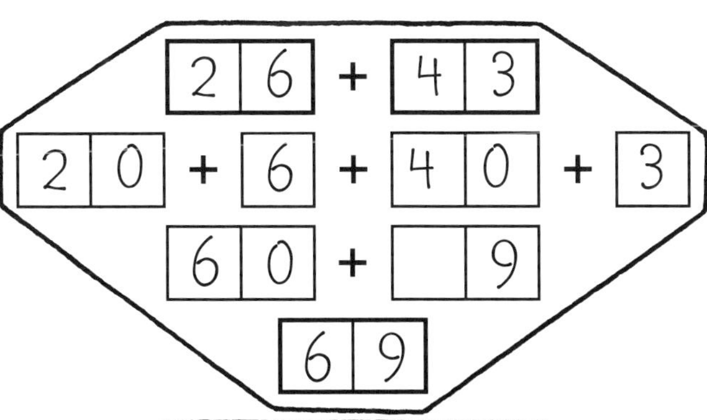


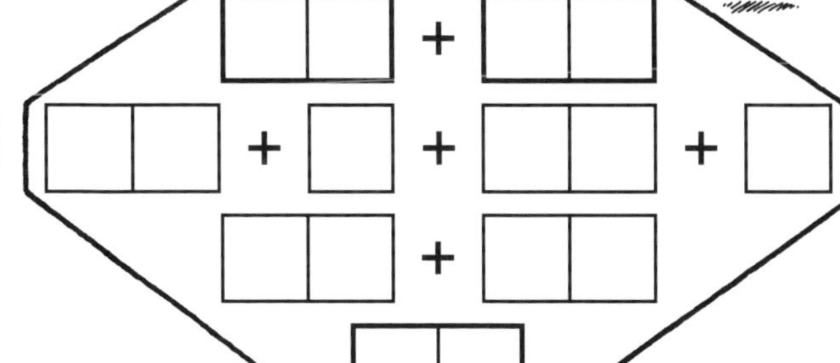

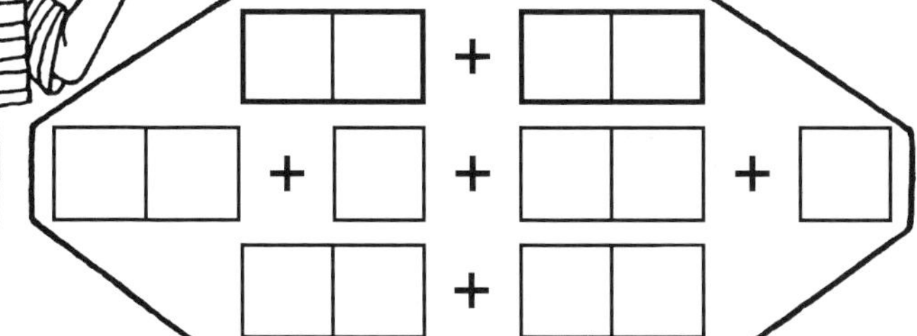

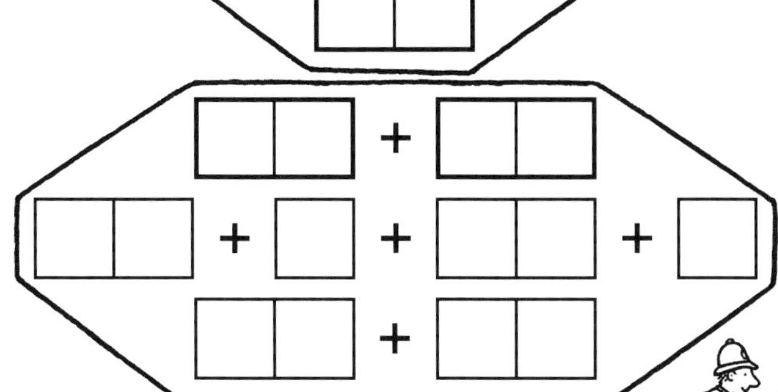

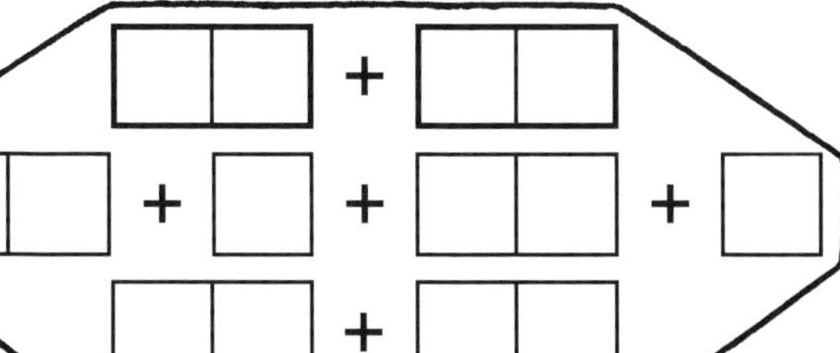

- **Write an addition sentence for each jewel.**

Example: 24 + 43 = 67

**Teachers' note** Before photocopying, write in the top row of each jewel a pair of two-digit numbers with a total of less than 100 and with units digits that will not cross the tens boundary. (Alternatively, children could select their own numbers with a total less than 100 from a list on the board.) Additions for more confident children could involve crossing tens boundaries.

# Loch Ness monsters

- **Find the difference. Count up from the smaller number.**

**25 – 17**

**21 – 18**

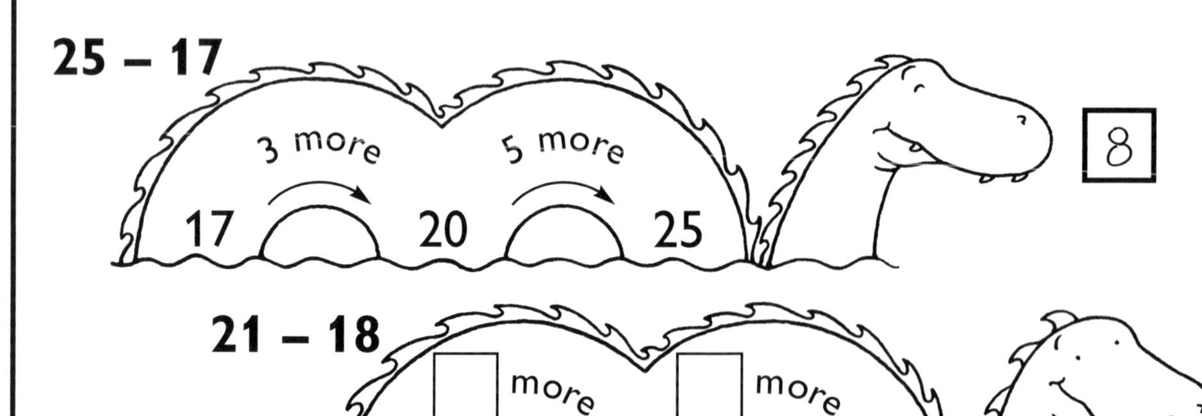


**104 – 95**

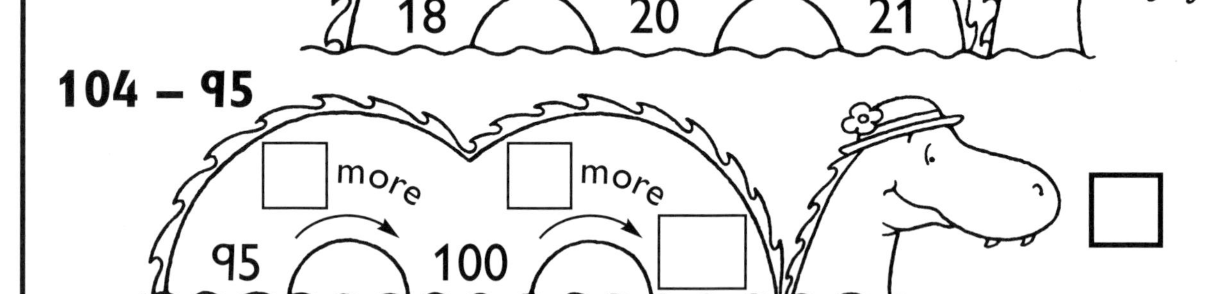


**24 – 19**

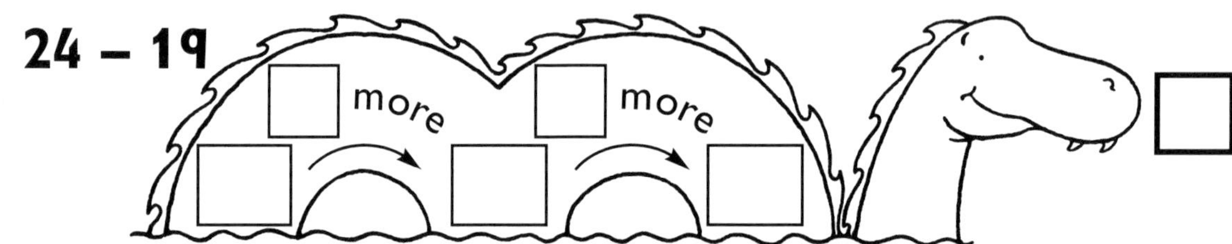


**33 – 26**

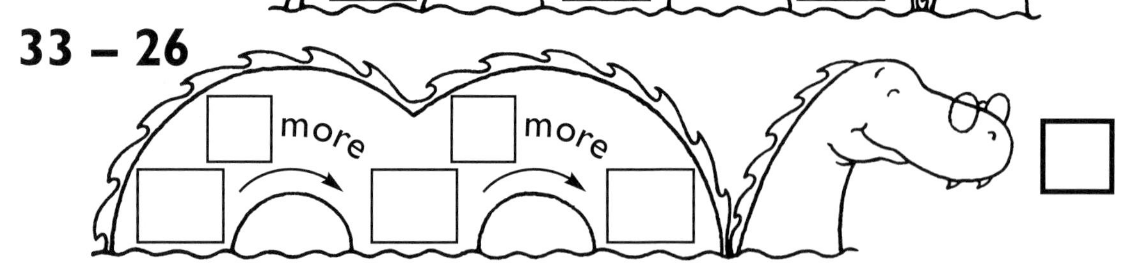

- **Answer these in the same way.**

34 – 28 = ☐ 41 – 35 = ☐ 55 – 48 = ☐

61 – 54 = ☐ 82 – 77 = ☐ 93 – 86 = ☐

102 – 98 = ☐ 103 – 95 = ☐ 103 – 96 = ☐

**Teachers' note** Begin the lesson by asking the children to count on in ones from a number in the nineties, up to and beyond 100. Then give pairs of numbers and ask the children to say the multiple of 10 or 100 that lies between them. Encourage the use of number bonds when counting up through a multiple of 10 or 100.

# Near doubles

• Find the total for each pair. Double one of the numbers, then add or subtract 1.

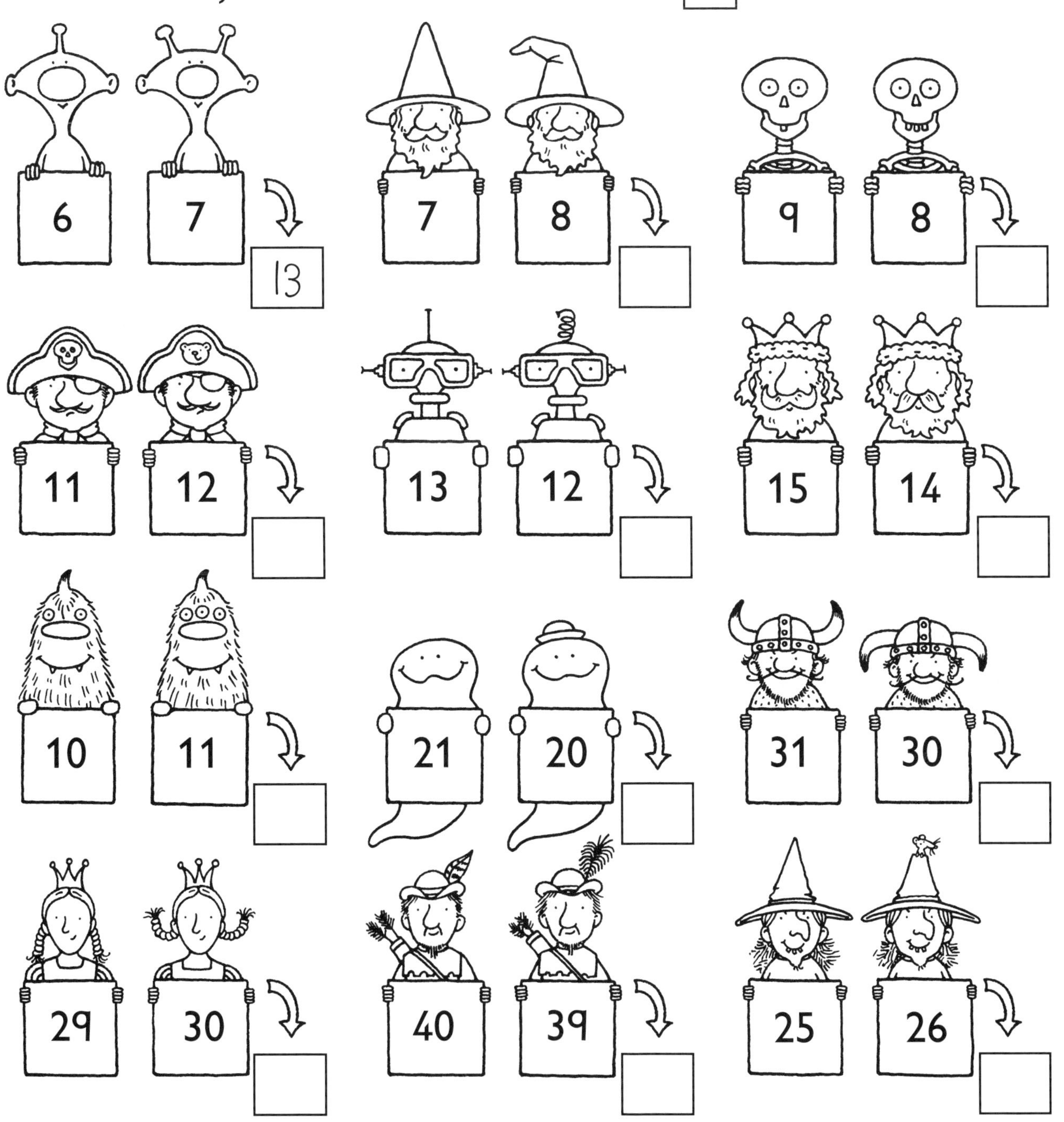

• Discuss with a partner how to answer these questions using doubles.

50 + 49    50 + 51    100 + 99    200 + 199

**Teachers' note** The children will need to know by heart, or be able to derive quickly, doubles of numbers to at least 15 and of some multiples of 10. The activity on page 42 could be used to revise these facts before starting this sheet. Encourage the children to explain each calculation strategy aloud to a partner. Use a range of vocabulary to reinforce addition and doubling words.

# Nine lives?

| 1 | 2 | 3 | 4 | 5 | 6 | 7 | 8 | 9 |
|---|---|---|---|---|---|---|---|---|
| 10 | 11 | 12 | | | 15 | 16 | 17 | 18 |
| 19 | 20 | | | | | | | |
| 28 | | | | | | | | |
| 37 | | | | | | | | |
| 46 | | | | | | | | |
| 55 | | | | | | | | |
| 64 | | | | | | | | |
| 73 | | | | | | | 80 | 81 |
| 82 | 83 | | | | | 88 | 89 | 90 |
| 91 | 92 | 93 | 94 | 95 | 96 | 97 | 98 | 99 |

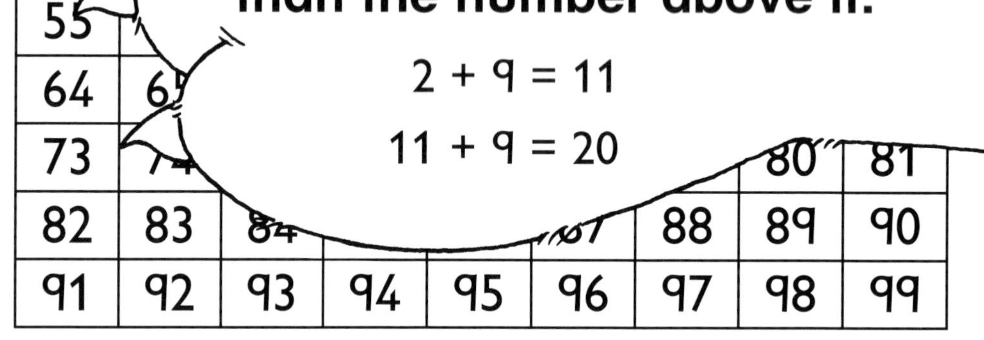


- **Complete these parts of the grid.**

**To add 9, you can add 10, then take away 1.**

| 3 |
|---|
| 12 |
| 21 |
| 30 |
| |
| |
| |
| |

| | 14 | |
|---|---|---|
| | 23 | 24 |
| | | |
| | | |
| | | |
| | | |
| | | |
| | | |

| 55 | 56 | | |
|---|---|---|---|
| 64 | | | |
| | | | |
| | | | |

| 58 | | | | | |
|---|---|---|---|---|---|
| | | | | | |
| | | | | | |
| | | | | | |

| 7 |
|---|
| 16 |
| |
| |
| |
| |
| |
| 70 |

- **Write all the numbers that come above 96 in the grid. Take away 9 each time.**

96, 87, ___, ___, ___, ___, ___, ___, ___, ___, ___

**Teachers' note** Ensure that the children are confident in adding and subtracting 10 to/from any two-digit number before tackling this activity. For the extension activity, remind the children that to subtract 9, they can subtract 10, then add 1.

# Crown adding

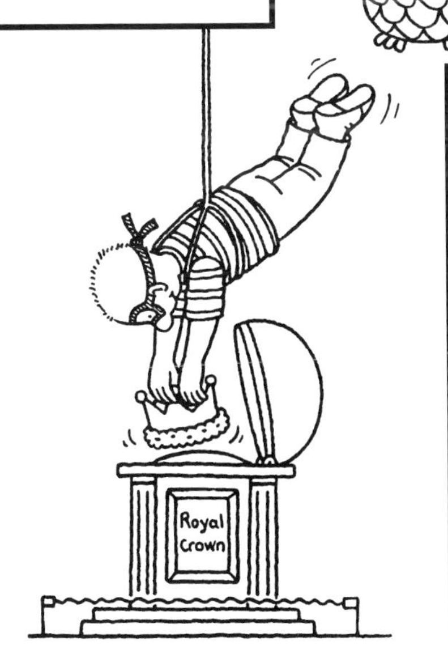


- **Use these crowns to add 19.**

**First add 20, then subtract 1.**

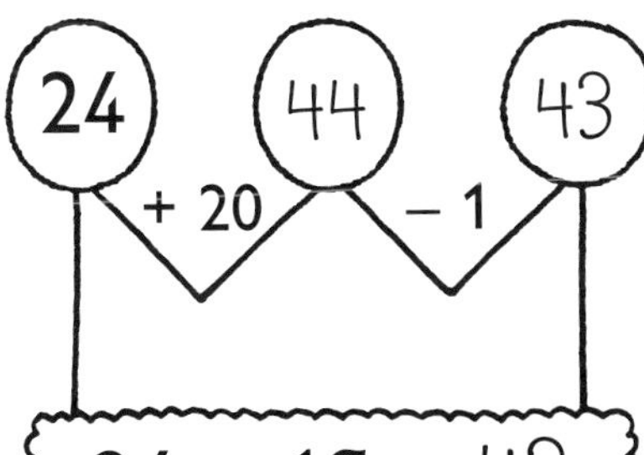


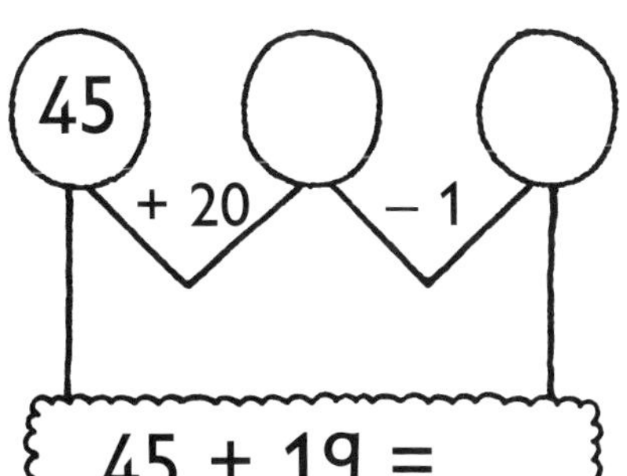


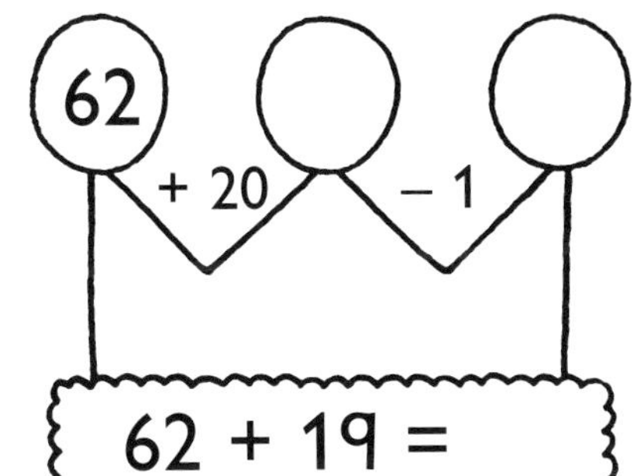


28 + 20 − 1

28 + 19 =

56 + 20 − 1

56 + 19 =

- **Use these crowns to add 21.**

**First add 20, then add 1.**

27 + 20 + 1

27 + 21 =

34 + 20 + 1

34 + 21 =

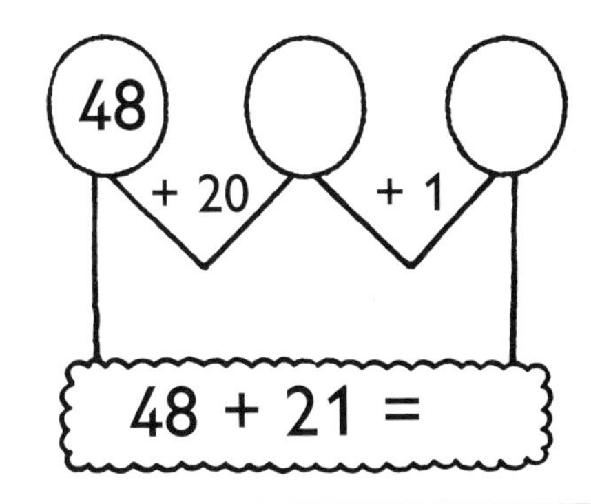

- **Answer these additions.**

| | | |
|---|---|---|
| 22 + 19 = ☐ | 55 + 19 = ☐ | 37 + 19 = ☐ |
| 54 + 19 = ☐ | 68 + 19 = ☐ | 44 + 19 = ☐ |
| 63 + 21 = ☐ | 57 + 21 = ☐ | 36 + 21 = ☐ |

**Teachers' note** Demonstrate how the crowns work: you start with the number in the jewel on the left, then add 20 and write the number in the central jewel, and then subtract 1 and write this in the third jewel. Ensure the children understand that this is the same as adding 19 to the number, and that the answer is the number in the third jewel. Also show how to use the crowns to add 21.

# Crown subtracting

- **Use these crowns to subtract 19.**

**First subtract 20, then add 1.**

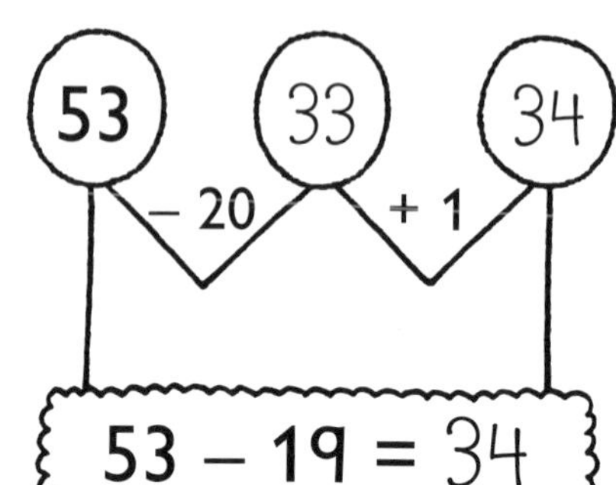


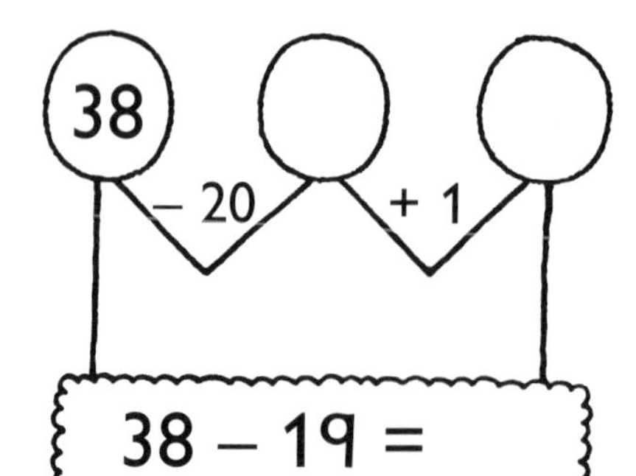

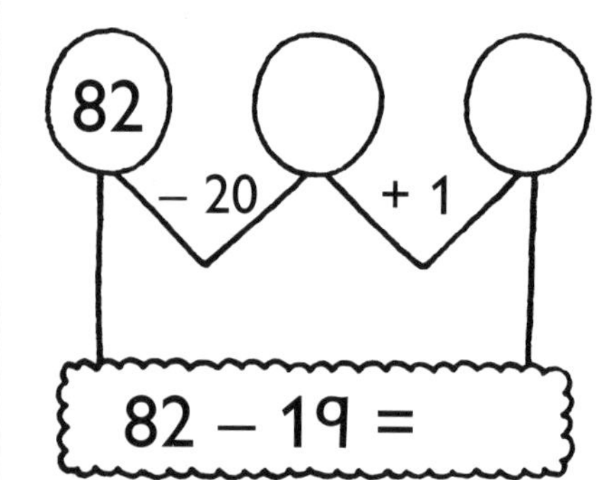


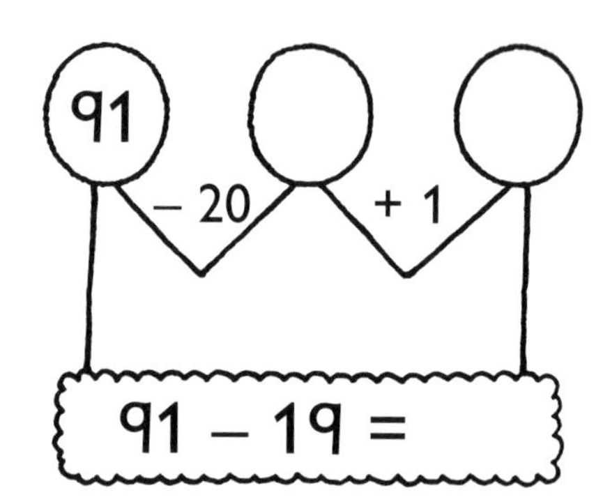


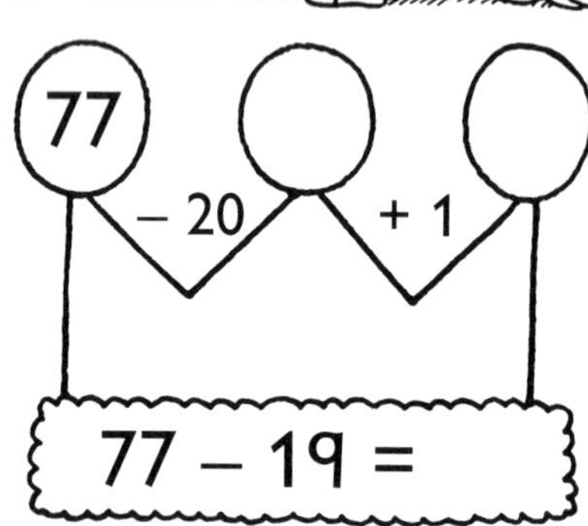


- **Use these crowns to subtract 21.**

**First subtract 20, then subtract 1.**

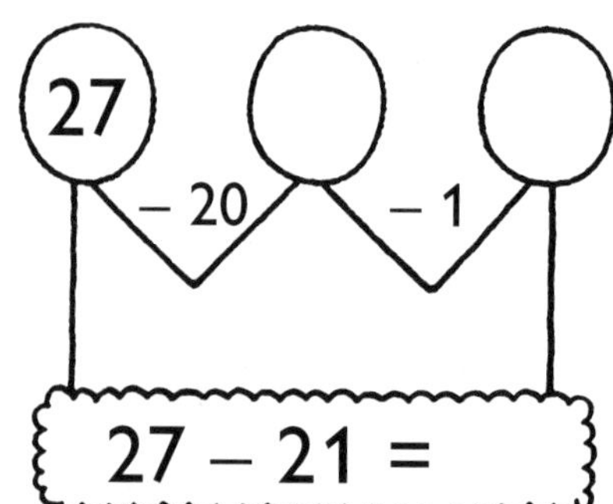


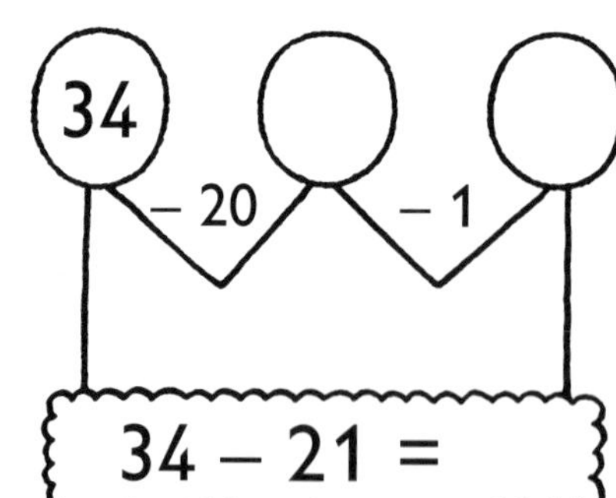


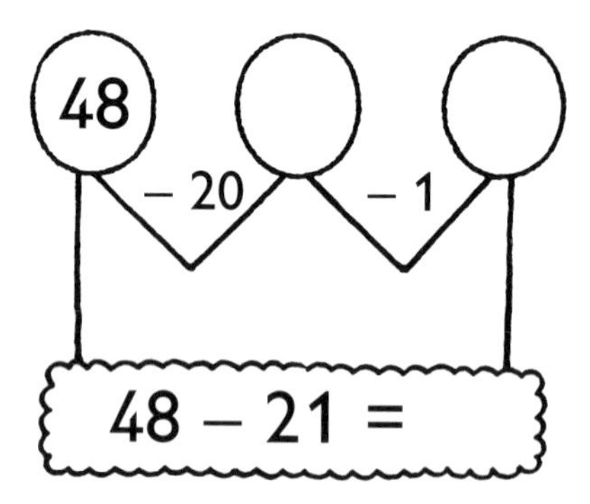

- **Answer these subtractions.**

62 – 19 = ☐ 85 – 19 = ☐ 97 – 19 = ☐

55 – 19 = ☐ 43 – 19 = ☐ 74 – 19 = ☐

66 – 21 = ☐ 82 – 21 = ☐ 99 – 21 = ☐

**Teachers' note** Demonstrate how the crowns work: you start with the number in the jewel on the left, then subtract 20 and write the number in the central jewel, and then add 1 and write this in the third jewel. Ensure the children understand that this is the same as subtracting 19 from the number, and that the answer is the number in the third jewel. Also show how to use the crowns to subtract 21.

# Cups and saucers

- **Use the facts on the cup to answer the question.**

3 + 6 = 9
13 + 6 = 19

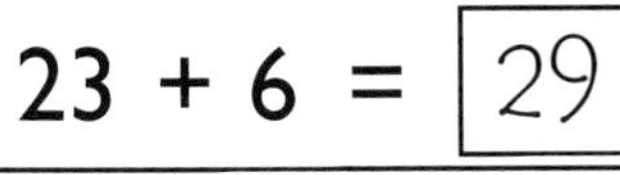

23 + 6 = 29

4 + 4 = 8
14 + 4 = 18

24 + ☐ = 28

2 + 5 = 7
12 + 5 = 17

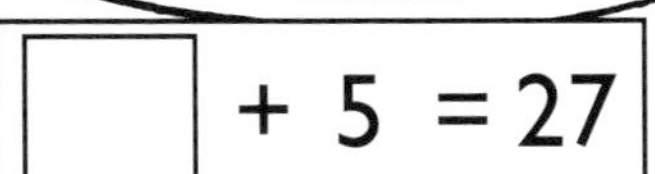

☐ + 5 = 27

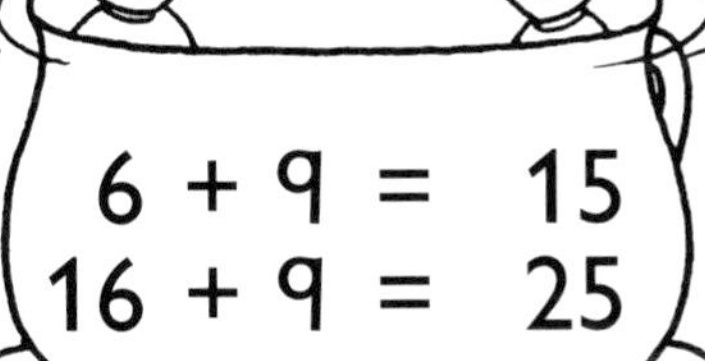

6 + 9 = 15
16 + 9 = 25

26 + 9 = ☐

7 + 9 = 16
17 + 9 = 26

27 + ☐ = 36

8 + 7 = 15
18 + 7 = 25

☐ + 7 = 35

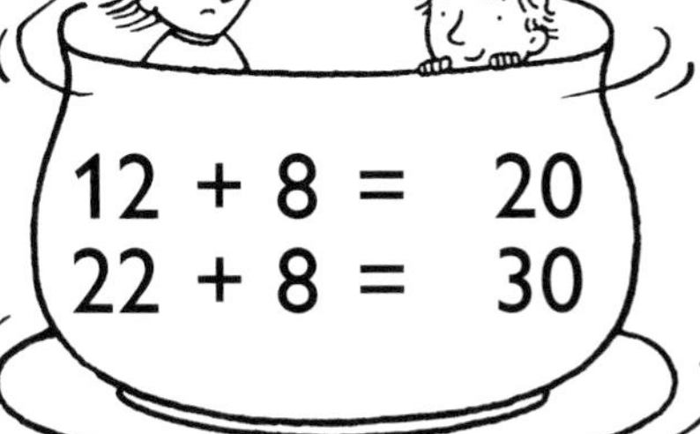

12 + 8 = 20
22 + 8 = 30

32 + 8 = ☐

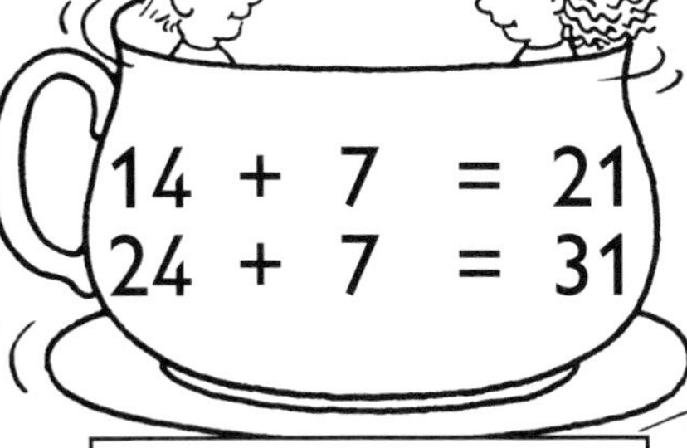

14 + 7 = 21
24 + 7 = 31

34 + ☐ = 41

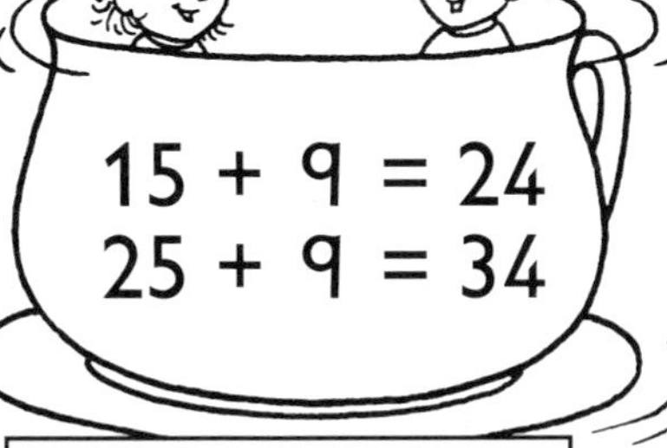

15 + 9 = 24
25 + 9 = 34

☐ + 9 = 44

- **Use facts you know to answer these.**

23 + 5 = ☐   41 + 4 = ☐   6 + 32 = ☐

22 + ☐ = 26   32 + ☐ = 35   4 + ☐ = 27

☐ + 5 = 37   ☐ + 4 = 29   ☐ + 42 = 49

**Teachers' note** In the extension activity, encourage the children to say aloud the facts they use, using a range of addition vocabulary (such as 'plus', 'add', 'makes', 'altogether', 'and', 'equals').

# Garden walls

- **Add the numbers next to each other in the wall. Write the total in the brick above.**
- **In the top row of each wall, ring the brick with the largest number.**

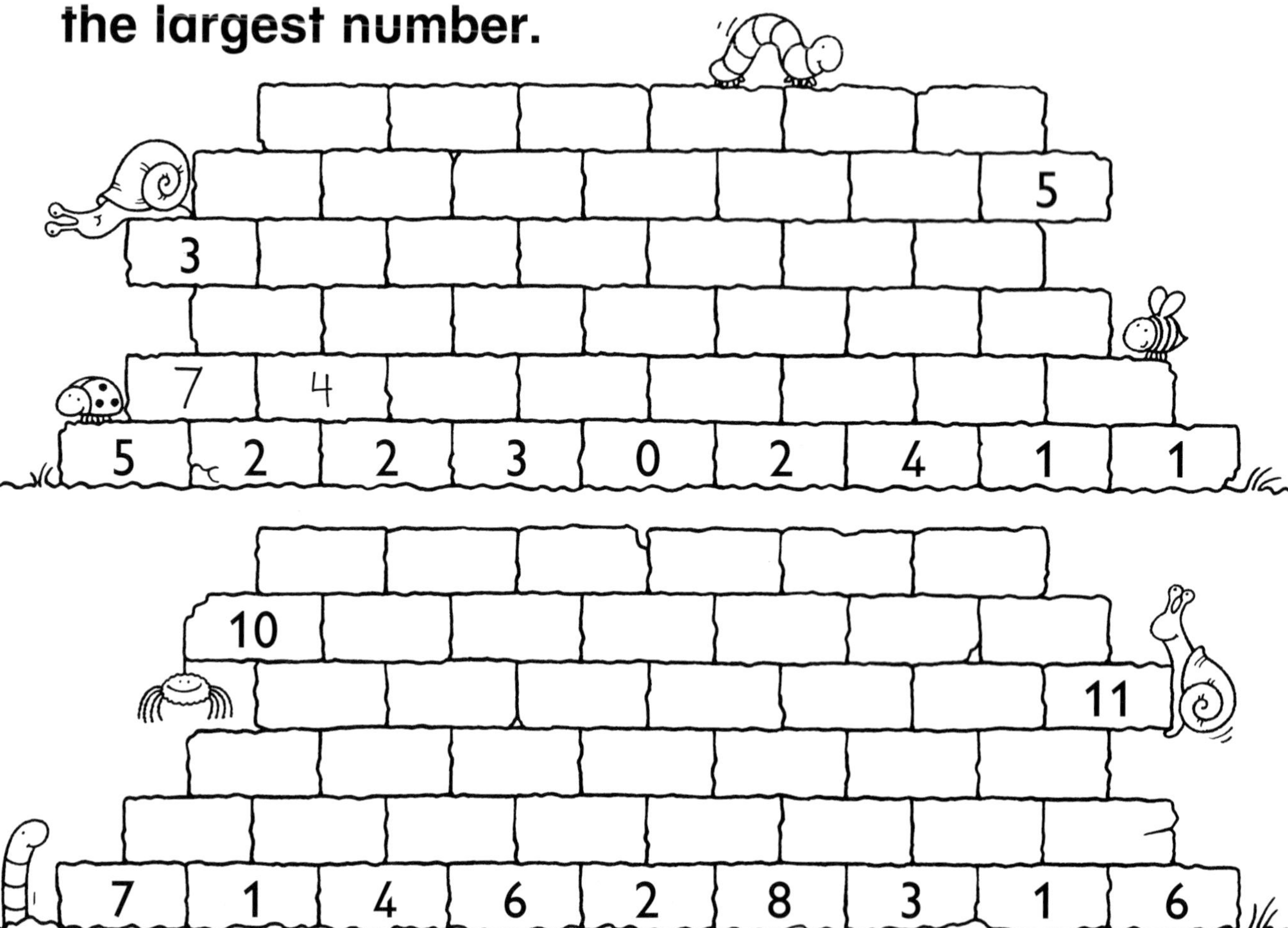

- **Answer these additions.**

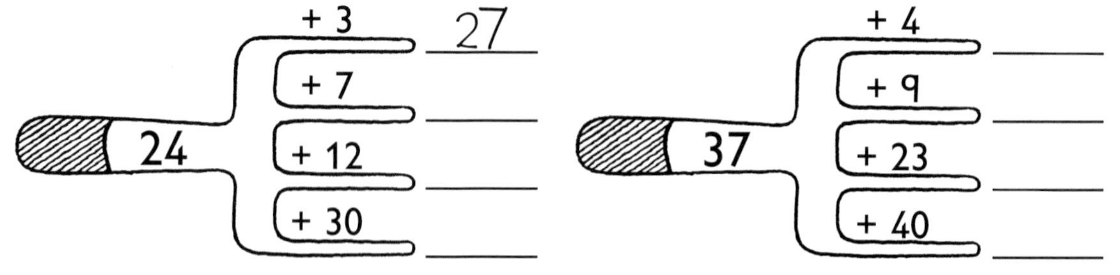


- **Now draw your own addition wall. Use some of these answers on the first row of bricks.**

**Teachers' note** Discuss that different strategies can be used: for example, adding 7 and 1 can be done by counting on 1 from 7 (the larger number), whereas adding 25 and 24 can be done using knowledge of place value, partitioning or using near doubles. You could fill in some numbers towards the top of the walls to help less confident children.

# Picture puzzles

**This code uses pictures to stand for numbers.**

38 27 46 59 68 49 86 77

2 3 5 4 6

- **These sums are written in code. Work out the answers in your head.**

1. − = 33
2. − = ____
3. − = ____
4. − = ____
5. − = ____
6. − = ____
7. − = ____
8. − = ____
9. − = ____
10. − = ____

- **Draw the missing pictures.**

− = 44

− = 72

− = 43

− = 65

− = 47

− = 82

**Teachers' note** The numbers in the code can be changed to provide differentiation. More confident children could be given larger single-digit numbers to subtract which involve crossing a tens boundary. Encourage the children to explain which number facts they used to help them answer the questions (for example, 'I know that 8 take away 5 is 3, so 68 take away 5 must be 63').

# Brain booster!

- **Work out the answers in your head.**

1. 38 – 16 = [22]
2. 25 – 14 = [ ]
3. 46 – 13 = [ ]
4. 43 – 12 = [ ]
5. 35 – 11 = [ ]
6. 49 – 17 = [ ]
7. 38 – 15 = [ ]

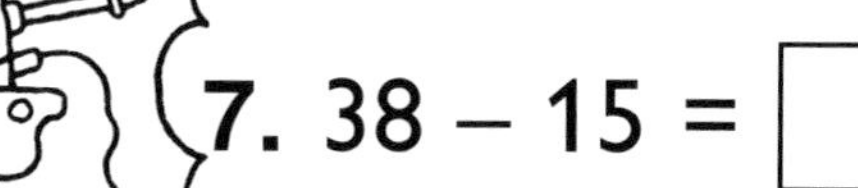

8. 58 – 17 = [ ]
9. 79 – 16 = [ ]



- **Explain to a partner how you worked out each answer.**

- **Make different questions with the answer [23]. Use these numbers.**

38 34 37 36 13 14 11 12 15

[ ] – ( ) = 23 [ ] – ( ) = 23

[ ] – ( ) = 23 [ ] – ( ) = 23

**Teachers' note** For further practice, mask the numbers and replace them with questions of a similar type, where numbers do not cross a tens boundary (for example, 89 – 18, 78 – 13, 95 – 12). Encourage the children to discuss how they worked out each answer (for example, '28 – 15: I know that 8 is 3 more than 5, so 28 is 3 more than 25, so 28 must be 13 more than 15').

# Super skier

☆ Put the cards in a pile, face down.
☆ Take the top card and write the number in the box. Answer the question.
☆ Put the card to the bottom of the pile.
☆ Take another card and continue.
☆ Try to reach **finish** as quickly as you can.

**Teachers' note** Each child will need a set of 1 to 9 digit cards. To make the activity easier, remove the cards greater than 5. To make it harder, remove the cards less than 5. If cards are not available, single-digit numbers can be written in before photocopying. As an extension, ask the children to write six subtractions with the answer 17.

# Bee careful counting

There are 100 bees in each beehive.

- **Write how many bees are in each row.**

 ______

 ______

______

 ______

 ______

- **Answer these addition questions.**

200 + 300 = ⬡ 100 + 500 = ⬡

400 + 500 = ⬡ 300 + 600 = ⬡

700 + 200 = ⬡ 400 + 400 = ⬡

800 + 200 = ⬡ 300 + 700 = ⬡

- **Answer these subtraction questions.**

700 – 300 = ⬡ 600 – 200 = ⬡

500 – 400 = ⬡ 800 – 300 = ⬡

700 – ⬡ = 200 1000 – ⬡ = 300

800 – ⬡ = 600 600 – ⬡ = 100

**Teachers' note** Encourage the children to check their answers using a related calculation: for example, by using number facts to 10 ('I know that 7 + 2 = 9, so 700 + 200 must be 900'). Additions could be checked using subtraction, and vice versa. Encourage the children to read the questions and answers aloud to develop appropriate vocabulary.

# At the shops: 1

**Teachers' note** Use these cards with the activity on page 36. They can also be used in a variety of other ways: for example, arranging the prices in order; picking pairs of cards and finding the total; working out change from 50p or £1; playing 'shop' and finding the correct coins with which to pay; and whole-class questions such as 'I have 10p. How much more do I need to buy this?'

# At the shops: 2

- Cut out the shopping cards from *At the shops: 1*.
- Put pairs of cards here. Find the total.

 +  = ?

- Write your additions here.

Example: 14p + 25p = 39p

__________ __________ __________

__________ __________ __________

__________ __________ __________

Now try this!

- Pick one card at a time. Find the change from 50p.

50p –  = ?

- Write the subtractions like this:

50p – 15p = 35p

**Teachers' note** The children will need copies of page 35 for this activity (pairs or small groups can share sets of the shopping cards). They could discuss the strategies used for finding totals and working out change from 50p. In the extension activity, the amount 50p could be masked and changed to £1 (or another amount) for more confident children.

# Sleeping snakes

- **For each start number, follow the arrows and write the finish number.**

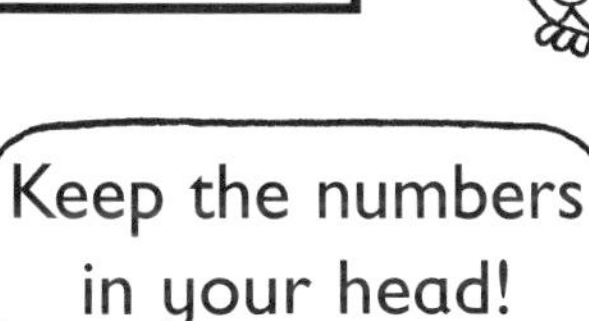


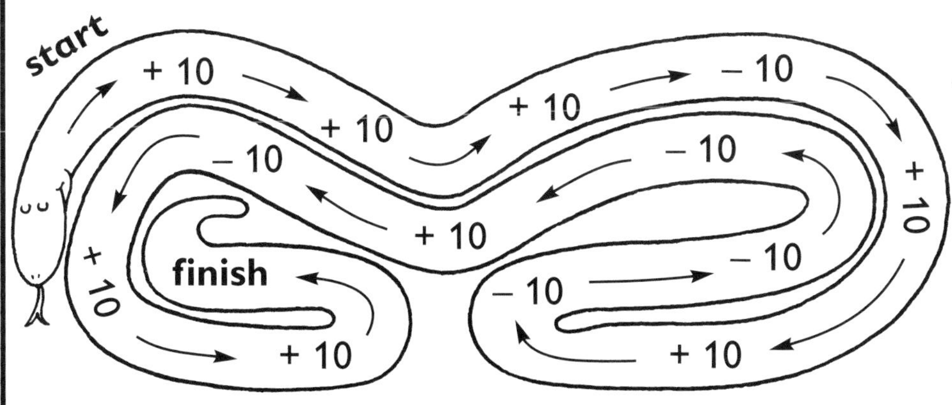


| start | finish |
|---|---|
| 50 | |
| 24 | |
| 41 | |

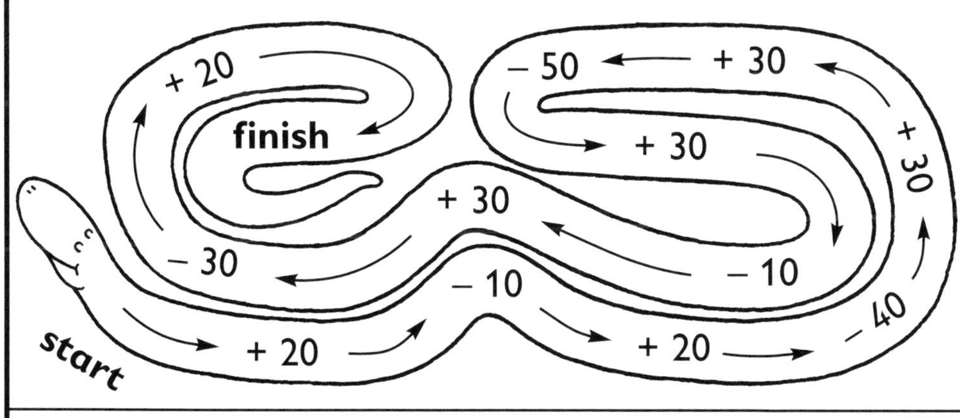


| start | finish |
|---|---|
| 30 | |
| 46 | |
| 27 | |

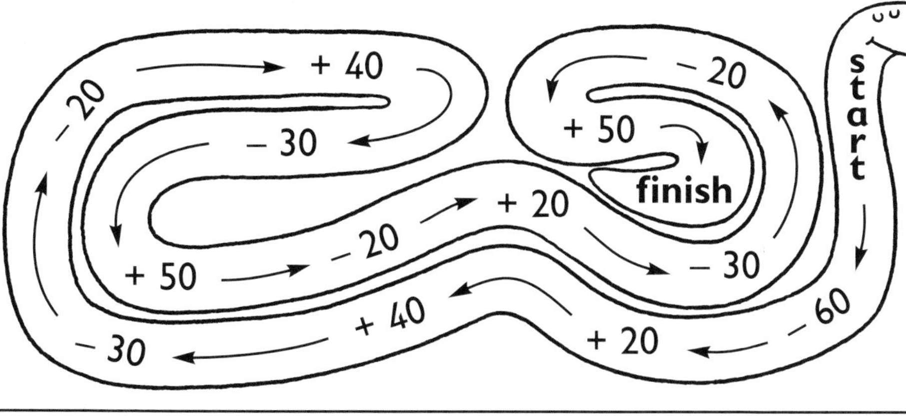


| start | finish |
|---|---|
| 76 | |
| 84 | |
| 69 | |

- **Choose another start number for each snake.**
- Predict **the finish number.**
- **Check by following the arrows.**

**Teachers' note** This page allows children to practise mentally adding and subtracting multiples of 10 to/from two-digit numbers, without crossing 100. Some children might spot a pattern in the finish numbers for each snake. If this happens, encourage them to predict the answers and then to follow the arrows as a means of checking the answers.

# Alien eyes

- **The aliens in each pair have the same number of eyes. How many eyes does each pair have?**

- **Now list the answers in order.**

2, ____________________

**The numbers above are in the** two times table.

Now try this!

- **Fill in the missing numbers.**

$2 \times 5 =$ ☐ $2 \times 4 =$ ☐ $2 \times 6 =$ ☐ $2 \times 3 =$ ☐

$2 \times 8 =$ ☐ $2 \times 2 =$ ☐ $2 \times 7 =$ ☐ $2 \times 1 =$ ☐

$2 \times 9 =$ ☐ $2 \times 10 =$ ☐

**Teachers' note** During the plenary, describe the questions using vocabulary such as: 'What is double 5? What is twice 5? How many is 5 add 5? What is two lots of 5? How many is 2 times 5? 2 multiplied by 5 equals…?' Invite the children to say how they worked out the answers or whether they have learned the doubles by heart.

# Alien legs

- **The aliens in each pair have the same number of legs. How many legs does one alien have?**

- **Use the answers above to complete these.**

6 ÷ 2 = ☐ 10 ÷ 2 = ☐ 2 ÷ 2 = ☐ 4 ÷ 2 = ☐

12 ÷ 2 = ☐ 16 ÷ 2 = ☐ 14 ÷ 2 = ☐ 8 ÷ 2 = ☐

20 ÷ 2 = ☐ 18 ÷ 2 = ☐

- **For each division fact above, write three statements.** Example:

**6 ÷ 2 = 3**

Half of **6** is **3**.
There are **3** twos in **6**.
**6** divided by two is **3**.

**Teachers' note** During the plenary, describe the questions using vocabulary such as: 'What is 10 divided by two? What is half 10? How many is 10 shared between two? How many twos in 10?' Invite the children to say how they worked out the answers or whether they have learned the halves by heart.

# Totalling tens

• **Count in tens. How many pounds in each row?**

£10 £10 £10 £10 £10 £10 £10 £10 £80

£10 £10 £10 £10 £10 £10 ____

£10 £10 £10 £10 £10 £10 £10 ____

£10 £10 £10 £10 ____

£10 £10 £10 £10 £10 £10 £10 £10 £10 ____

£10 £10 £10 £10 £10 ____

• **Write the answers to these questions.**

**1.** 3 lots of 10 = ☐　　**2.** 5 lots of 10 = ☐

**3.** $4 \times 10 =$ ☐　　**4.** $6 \times 10 =$ ☐

**5.** 10 multiplied by 4 = ☐　　**6.** 10 multiplied by 7 = ☐

**7.** $10 \times 2 =$ ☐　　**8.** $10 \times 9 =$ ☐

**9.** 6 times 10 = ☐　　**10.** 10 times 8 = ☐

• **Complete the division facts for the** ten times table **as quickly as you can.**

$10 \div 10 =$ ☐　　☐ $\div 10 = 6$　　$0 \div 10 =$ ☐

$30 \div$ ☐ $= 3$　　$40 \div 10 =$ ☐　　☐ $\div 10 = 8$

$50 \div 10 =$ ☐　　☐ $\div 10 = 9$　　$70 \div 10 =$ ☐

$20 \div$ ☐ $= 2$　　$100 \div 10 =$ ☐

**Teachers' note** Encourage the children to see the link between counting in tens, the multiplication facts and the division facts. During the plenary, use a range of vocabulary to describe the facts and invite the children to read the questions and answers aloud.

# Rows of stickers

**Stickers come in rows of five.**

- **Count in fives. How many stickers in each set?**

10

- **Write the answers to these questions.**

**1.** 3 rows of 5 = ☐

**2.** 2 rows of 5 = ☐

**3.** $5 \times 5 =$ ☐

**4.** $6 \times 5 =$ ☐

**5.** 5 multiplied by 2 = ☐

**6.** 5 multiplied by 5 = ☐

**7.** $5 \times 4 =$ ☐

**8.** $5 \times 6 =$ ☐

**9.** 5 times 4 = ☐

**10.** 3 times 5 = ☐

Now try this!

- **Complete the division facts for the** five times table **as quickly as you can.**

$15 \div 5 =$ ☐ $\quad 10 \div 5 =$ ☐ $\quad$ ☐ $\div 5 = 0$

☐ $\div 5 = 1$ $\quad 20 \div 5 =$ ☐ $\quad 25 \div$ ☐ $= 5$

$30 \div 5 =$ ☐ $\quad$ ☐ $\div 5 = 10$

**Teachers' note** Encourage the children to see the link between counting in fives, the multiplication facts and the division facts. During the plenary, use a range of vocabulary to describe the facts and invite the children to read the questions and answers aloud. Discuss that $3 \times 5$ has the same answer as $5 \times 3$ by showing that 15 stickers can be seen as 3 rows of 5 or 5 columns of 3.

# Eat your greens game

- **Play this game with a partner.**

**You need** a small counter each and one dice.

☆ Choose who will take hutch 1 and who will take hutch 2.

☆ Take turns to roll the dice. Move your counter.

☆ Double the number you land on. If you can find the answer on your hutch, cross it off.

☆ Continue until you both reach **finish**. Whoever crosses off the most numbers is the winner.

25

11 35 12 14 40 13 30 45 13 50

25

12 20 45 7 15 35 9 14 8

finish

**hutch 1**

| | | | | |
|---|---|---|---|---|
| 2 | 26 | 18 | 16 | 4 |
| 20 | 28 | 24 | 6 | 22 |
| 10 | 100 | 40 | 30 | 50 |
| 80 | 70 | 90 | 14 | 12 |

**hutch 2**

| | | | | |
|---|---|---|---|---|
| 60 | 100 | 26 | 8 | 14 |
| 24 | 16 | 2 | 30 | 28 |
| 4 | 90 | 20 | 50 | 10 |
| 18 | 22 | 60 | 80 | 70 |

**Teachers' note** Each pair needs one copy of this sheet. The game practises recalling or deriving doubles of numbers to 15 and of multiples of 5 to 50. Discuss strategies for doubling numbers such as 15 and 35: for example, by doubling 10 or 30, then doubling 5 and adding the two parts.

# Helter skelter game

- **Play this game with a partner.**

**You need** a small counter each and one dice.

☆ Choose who will take mat 1 and who will take mat 2.

☆ Take turns to roll the dice. Move your counter.

☆ Halve the number you land on. If you can find the answer on your mat, cross it off.

☆ Continue until you both reach **finish**. Whoever crosses off the most numbers is the winner.

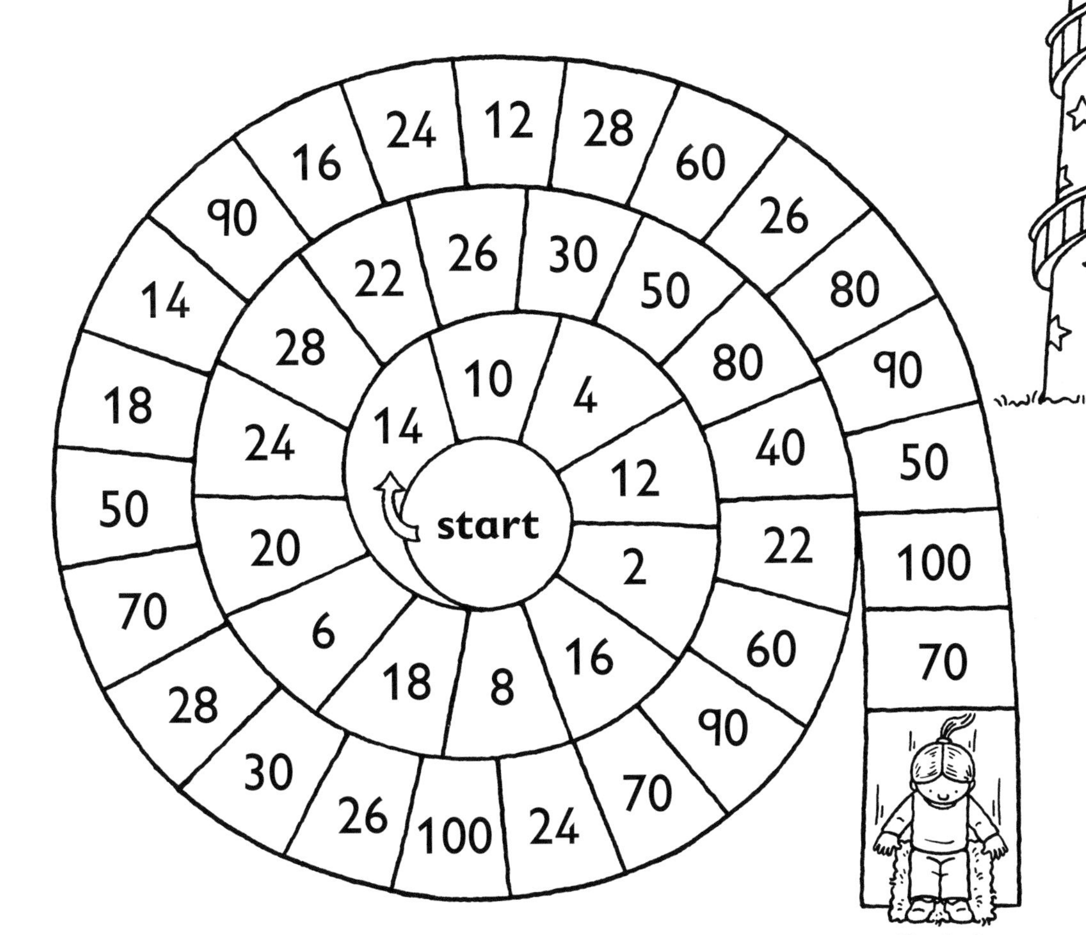


**finish**

**mat 1**

| | | | | |
|---|---|---|---|---|
| 1 | 13 | 8 | 9 | 2 |
| 10 | 14 | 12 | 3 | 11 |
| 5 | 50 | 20 | 15 | 25 |
| 40 | 35 | 45 | 7 | 6 |

**mat 2**

| | | | | |
|---|---|---|---|---|
| 30 | 50 | 13 | 4 | 7 |
| 12 | 8 | 1 | 15 | 14 |
| 2 | 45 | 10 | 25 | 5 |
| 9 | 11 | 30 | 40 | 35 |

**Teachers' note** Each pair needs one copy of this sheet. The game practises recalling or deriving halves of even numbers to 30 and of multiples of 10 to 100. Discuss strategies for halving numbers such as 30 and 90: for example, by halving 20 or 80, then halving 10 and adding the two parts.

# In the dinner hall

## • Answer these multiplication questions.

4 tables each have 1 plate.
How many plates?
4 lots of 1

$4 \times 1 =$ 4

3 tables each have 2 bowls.
How many bowls?
3 lots of 2

$3 \times 2 =$ ☐

5 tables each have 2 plates.
How many plates?
5 lots of 2

$5 \times 2 =$ ☐

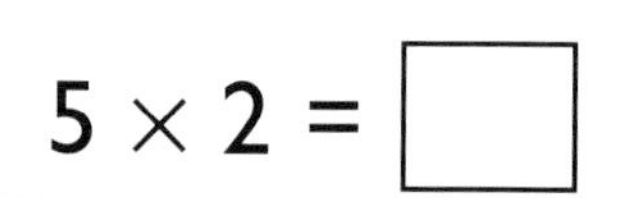

7 tables each have 5 bowls.
How many bowls?
7 lots of 5

$7 \times 5 =$ ☐

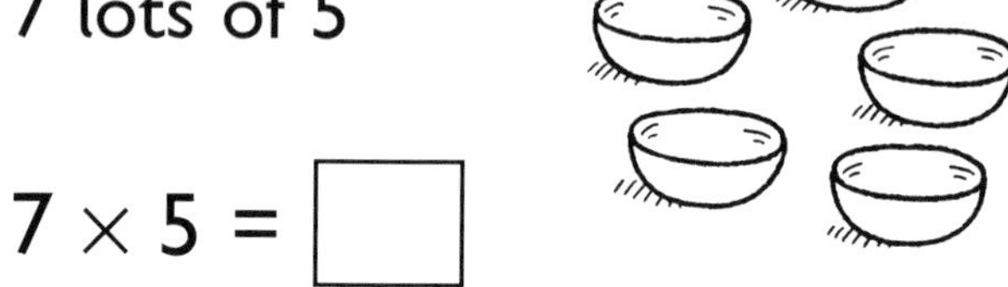

6 tables each have 5 spoons.
How many spoons?
6 lots of 5

$6 \times 5 =$ ☐

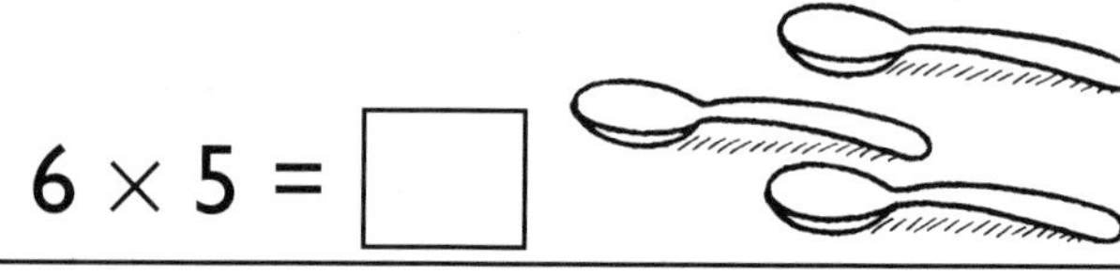

8 tables each have 2 forks.
How many forks?
8 lots of 2

$8 \times 2 =$ ☐

9 tables each have 5 knives.
How many knives?
9 lots of 5

$9 \times 5 =$ ☐

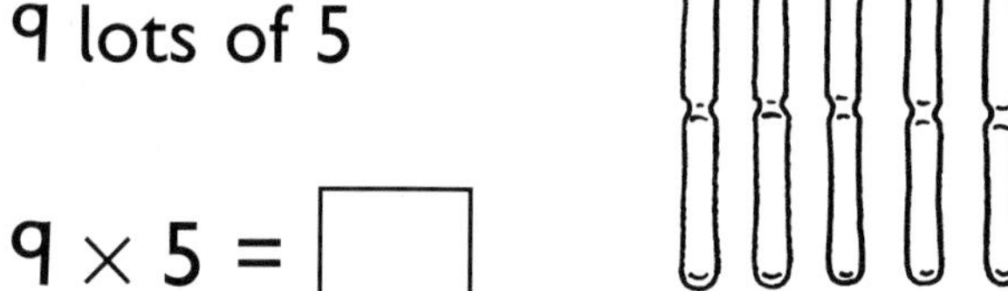

2 tables each have 5 cups.
How many cups?
2 lots of 5

$2 \times 5 =$ ☐

Now try this!

## • How many tables are needed?

**Two** items are put on each table.

18 items split into **twos**. → $18 \div 2 =$ 9 $6 \div 2 =$ ☐ $20 \div 2 =$ ☐

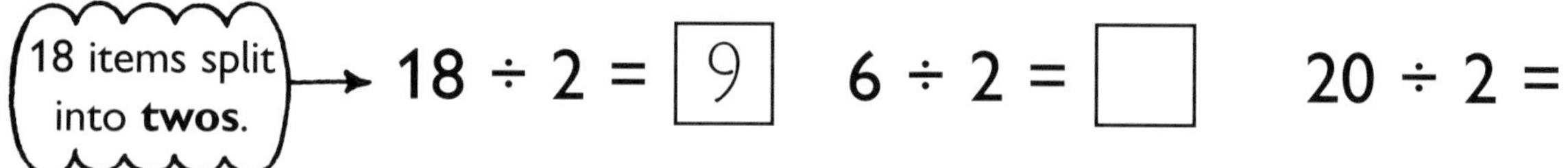

**Five** items are put on each table.

20 items split into **fives**. → $20 \div 5 =$ ☐ $50 \div 5 =$ ☐ $35 \div 5 =$ ☐

**Teachers' note** This multiplication approach uses the idea of 'lots of'. The division approach is the inverse of 'lots of' and involves grouping (rather than sharing). Introduce the grouping approach by giving a child a number of pencils and asking him or her to put one on each table. Ask: 'How many tables are needed?' Repeat for groups of two and then five.

# Lots for dinner

## • Answer these multiplication questions.

4 tables each have 10 plates.
How many plates?
4 lots of 10

$4 \times 10 =$ 40

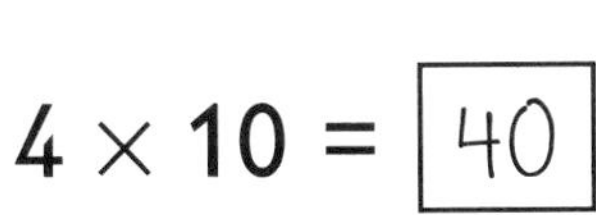

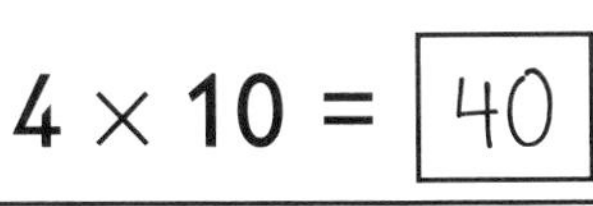

3 tables each have 10 bowls.
How many bowls?
3 lots of 10

$3 \times 10 =$ ☐

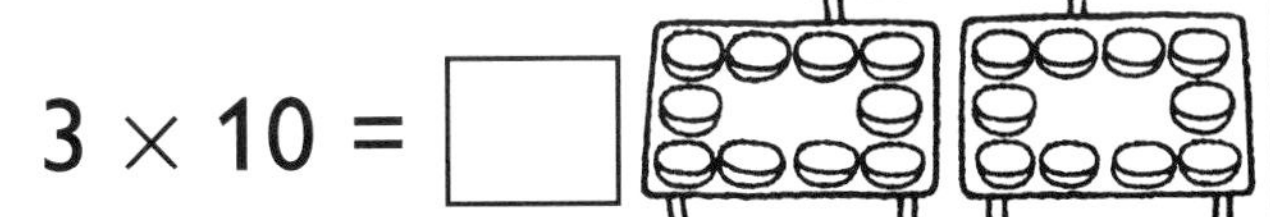

5 tables each have 10 plates.
How many plates?
5 lots of 10

$5 \times 10 =$ ☐

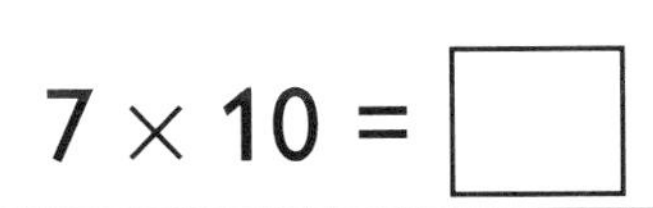

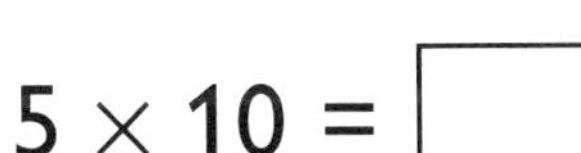

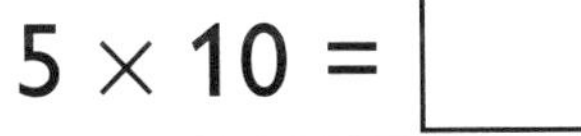

7 tables each have 10 bowls.
How many bowls?
7 lots of 10

$7 \times 10 =$ ☐

6 tables each have 10 spoons.
How many spoons?
6 lots of 10

$6 \times 10 =$ ☐

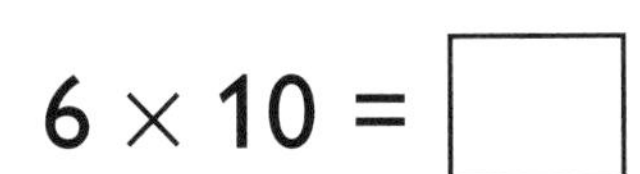

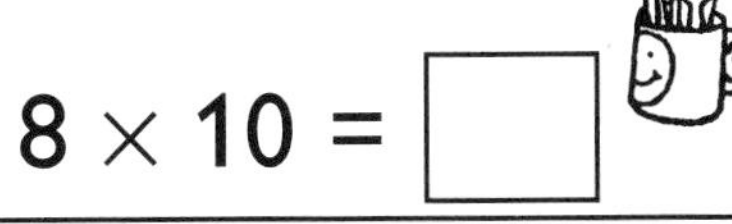

8 tables each have 10 forks.
How many forks?
8 lots of 10

$8 \times 10 =$ ☐

9 tables each have 10 knives.
How many knives?
9 lots of 10

$9 \times 10 =$ ☐

2 tables each have 10 cups.
How many cups?
2 lots of 10

$2 \times 10 =$ ☐

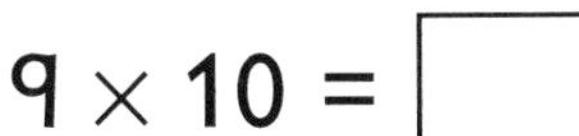

Now try this!

Ten items are put on each table.

**• How many tables are needed?**

90 items split into **tens**.

| | | |
|---|---|---|
| $90 \div 10 =$ 9 | $40 \div 10 =$ ☐ | $50 \div 10 =$ ☐ |
| $20 \div 10 =$ ☐ | $80 \div 10 =$ ☐ | $30 \div 10 =$ ☐ |
| $70 \div 10 =$ ☐ | $60 \div 10 =$ ☐ | $10 \div 10 =$ ☐ |

**Teachers' note** This multiplication approach uses the idea of 'lots of'. The division approach is the inverse of 'lots of' and involves grouping (rather than sharing). Introduce the grouping approach by giving a child 20 or 30 pencils and asking him or her to put 10 on each table. Ask: 'How many tables are needed?'

# Array of sunshine

- **Draw smiley faces. Then fill in the answer.**

| 2 rows of 4 | 3 rows of 3 | 2 rows of 5 |
|---|---|---|
| 😊😊😊😊<br>😊😊😊😊<br>$2 \times 4 =$ 8 | $3 \times 3 =$ ☐ | $2 \times 5 =$ ☐ |
| 3 rows of 4 | 4 rows of 5 | 3 rows of 2 |
| $3 \times 4 =$ ☐ | $4 \times 5 =$ ☐ | $3 \times 2 =$ ☐ |
| 5 rows of 3 | 4 rows of 4 | 5 rows of 5 |
| $5 \times 3 =$ ☐ | $4 \times 4 =$ ☐ | $5 \times 5 =$ ☐ |

Now try this!

- **Write the answers as quickly as you can.**

| | | | |
|---|---|---|---|
| $2 \times 2 =$ 4 | $2 \times 4 =$ ☐ | $3 \times 3 =$ ☐ | $4 \times 3 =$ ☐ |
| $2 \times 5 =$ ☐ | $4 \times 4 =$ ☐ | $3 \times 4 =$ ☐ | $2 \times 3 =$ ☐ |
| $4 \times 5 =$ ☐ | $3 \times 2 =$ ☐ | $5 \times 5 =$ ☐ | $5 \times 4 =$ ☐ |
| $3 \times 5 =$ ☐ | $4 \times 2 =$ ☐ | $5 \times 2 =$ ☐ | $5 \times 3 =$ ☐ |

**Teachers' note** Before the children attempt the extension activity, discuss that arrays can be described as rows or as columns, so 3 rows of 4 can also be described as 4 columns of 3. Thus, $3 \times 4$ has the same answer as $4 \times 3$. At the end of the lesson, encourage the children to say the questions and answers aloud using a range of vocabulary.

# Answers

**p 6**

1. 3, 8, 13, 18, 23, 28, 33, 38, 43, 48, 53, 58, 63, 68, 73, 78, 83, 88, 93, 98
2. 1, 6, 11, 16, 21, 26, 31, 36, 41, 46, 51, 56, 61, 66, 71, 76, 81, 86, 91, 96
3. 0, 5, 10, 15, 20, 25, 30, 35, 40, 45, 50, 55, 60, 65, 70, 75, 80, 85, 90, 95, 100

**Now try this!**
97, 92, 87, 82, 77, 72, 67, 62, 57, 52, 47, 42, 37, 32, 27, 22, 17, 12, 7, 2

**p 7**
8, 10, 12, 14, 16, 18, 20, 22, 24

**p 8**

1. 2, 12, 22, 32, 42, 52, 62, 72, 82, 92
2. 6, 16, 26, 36, 46, 56, 66, 76, 86, 96
3. 8, 18, 28, 38, 48, 58, 68, 78, 88, 98

**Now try this!**
94, 84, 74, 64, 54, 44, 34, 24, 14, 4
97, 87, 77, 67, 57, 47, 37, 27, 17, 7
99, 89, 79, 69, 59, 49, 39, 29, 19, 9

**p 9**

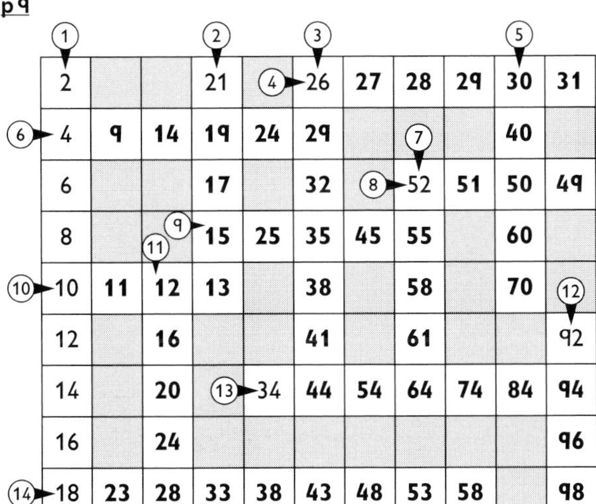

| 2 | | | 21 | | 26 | 27 | 28 | 29 | 30 | 31 |
|---|---|---|---|---|---|---|---|---|---|---|
| 4 | 9 | 14 | 19 | 24 | 29 | | | | 40 | |
| 6 | | | 17 | | 32 | | 52 | 51 | 50 | 49 |
| 8 | | | 15 | 25 | 35 | 45 | 55 | | 60 | |
| 10 | 11 | 12 | 13 | | 38 | | 58 | | 70 | |
| 12 | | 16 | | | 41 | | 61 | | | 92 |
| 14 | | 20 | | 34 | 44 | 54 | 64 | 74 | 84 | 94 |
| 16 | | 24 | | | | | | | | 96 |
| 18 | 23 | 28 | 33 | 38 | 43 | 48 | 53 | 58 | | 98 |

**p 11**
**Now try this!**
30, 41, 49, 64, 69, 77, 93, 96

**p 14**

| 1 | 2 | 3 | 4 | 5 | 6 | 7 | 8 | 9 | 10 |
|---|---|---|---|---|---|---|---|---|---|
| 11 | 13 | 12 | 14 | 15 | 16 | 17 | 18 | 20 | 19 |
| 21 | 22 | 23 | 24 | 35 | 26 | 27 | 28 | 29 | 30 |
| 31 | 32 | 33 | 34 | 25 | 36 | 37 | 38 | 49 | 40 |
| 41 | 42 | 52 | 44 | 45 | 46 | 47 | 48 | 39 | 50 |
| 51 | 43 | 53 | 54 | 56 | 55 | 57 | 58 | 59 | 60 |
| 61 | 62 | 63 | 64 | 65 | 66 | 67 | 68 | 69 | 80 |
| 71 | 81 | 73 | 74 | 75 | 76 | 77 | 78 | 79 | 70 |
| 72 | 82 | 83 | 84 | 86 | 85 | 87 | 99 | 89 | 90 |
| 91 | 92 | 93 | 94 | 95 | 96 | 97 | 98 | 88 | 100 |

**Now try this!**
13 (25) 22 (18) 29
37 44 45 (54) (53)
27 (72) 35 (28) 86 95
52 58 88 (96) (95) 99

(18) (17) 19 20 24
43 52 (85) (76) 95
66 (77) 69 75 76 (68)
34 81 (87) 86 (82) 88

**p 18**
**Now try this!**
85 + 15 = 100
25 + 75 = 100
5 + 95 = 100
65 + 35 = 100
55 + 45 = 100

**p 19**

| $^{1}$2 | 6 | | $^{2}$6 | 6 |
|---|---|---|---|---|
| 8 | | $^{3}$4 | 9 | |
| | $^{4}$1 | 3 | | $^{5}$9 |
| $^{6}$7 | 2 | | $^{7}$6 | 5 |
| 4 | | $^{8}$8 | 3 | |

| $^{1}$4 | 2 | | $^{2}$5 | 8 |
|---|---|---|---|---|
| 7 | | $^{3}$3 | 1 | |
| | $^{4}$6 | 4 | | $^{5}$7 |
| $^{6}$8 | 9 | | $^{7}$5 | 5 |
| 3 | | $^{8}$2 | 4 | |

| $^{1}$9 | 5 | | $^{2}$6 | 4 |
|---|---|---|---|---|
| 2 | | $^{3}$6 | 3 | |
| | $^{4}$5 | 8 | | $^{5}$9 |
| $^{6}$7 | 4 | | $^{7}$8 | 1 |
| 2 | | $^{8}$7 | 3 | |

**p 22**
23 24
22 24
44 37

**Now try this!**
43 39 54
57 52 55
63 66 73

**p 24**
8
3
9
5
7

**Now try this!**
6 6 7
7 5 7
4 8 7

**p 26**

| 3 |
|---|
| 12 |
| 21 |
| 30 |
| 39 |
| 48 |
| 57 |
| 66 |

| | 14 | |
|---|---|---|
| | 23 | 24 |
| | 32 | 33 |
| | 41 | |
| 49 | 50 | |
| | 59 | |
| | 68 | |
| | 77 | |

| 55 | 56 | | |
|---|---|---|---|
| 64 | 65 | 66 | |
| 73 | 74 | 75 | 76 |
| | | 84 | |

| 58 | 59 | | | | |
|---|---|---|---|---|---|
| 67 | 68 | 69 | | | |
| | | 78 | 79 | 80 | 81 |
| | | | | 89 | 90 |

| 7 |
|---|
| 16 |
| 25 |
| 34 |
| 43 |
| 52 |
| 61 |
| 70 |

**Now try this!**
96, 87, 78, 69, 60, 51, 42, 33, 24, 15, 6

**p 27**
43 64
81 47 75
48 55 69

**Now try this!**
41 74 56
73 87 63
84 78 57

**p 28**
34 19
63 72 58
6 13 27

**Now try this!**
43 66 78
36 24 55
45 61 78

**p 29**

29 4 22
35 9 28
40 7 35

**Now try this!**

28 45 38
4 3 23
32 25 7

**p 30**

60 67 56 58 (69) 42
23 37 30 26 32 37 5
3 20 17 13 13 19 18
11 9 8 5 8 11 7
7 4 5 3 2 6 5 2
5 2 2 3 0 2 4 1 1

71 130 144 (150) 137 99
10 61 69 75 75 62 37
28 33 36 39 36 26 11
13 15 18 18 21 15 11
8 5 10 8 10 11 4 7
7 1 4 6 2 8 3 1 6

**Now try this!**

27 41
31 46
36 60
54 77

**p 31**

**1.** 33 **2.** 42
**3.** 56 **4.** 81
**5.** 25 **6.** 46
**7.** 74 **8.** 36
**9.** 80 **10.** 64

**Now try this!**

lips web
world plane
car – lips flower – dog

**p 32**

**1.** 22 **2.** 11 **3.** 33
**4.** 31 **5.** 24
**6.** 32 **7.** 23
**8.** 41 **9.** 63

**Now try this!**

38 – 15 34 – 11
37 – 14 36 – 13

**p 37**

| start | finish |
|---|---|
| 50 | **80** |
| 24 | **54** |
| 41 | **71** |

| start | finish |
|---|---|
| 30 | **70** |
| 46 | **86** |
| 27 | **67** |

| start | finish |
|---|---|
| 76 | **86** |
| 84 | **94** |
| 69 | **79** |